Topical Conference: Waste Plastics

AA002657

Held at the 2024 AIChE Annual Meeting

San Diego, California, USA
27-31 October 2024

ISBN: 979-8-3313-1672-3

Topical Conference: Waste Plastics

Held at the 2024 AIChE Annual Meeting

San Diego, California, USA
27-31 October 2024

AIChE
The Global Home of Chemical Engineers

Printed from e-media with permission by:

Curran Associates, Inc.
57 Morehouse Lane
Red Hook, NY 12571

CURRAN ASSOCIATES INC. proceedings.com

Some format issues inherent in the e-media version may also appear in this print version.

Copyright© (2024) by AIChE
All rights reserved.

Printed with permission by Curran Associates, Inc. (2025)

For permission requests, please contact AIChE
at the address below.

AIChE
120 Wall Street, FL 23
New York, NY 10005-4020

Phone: (800) 242-4363
Fax: (203) 775-5177

www.aiche.org

Additional copies of this publication are available from:

Curran Associates, Inc.
57 Morehouse Lane
Red Hook, NY 12571 USA
Phone: 845-758-0400
Fax: 845-758-2633
Email: curran@proceedings.com
Web: www.proceedings.com

TABLE OF CONTENTS

ANALYSIS AND ASSESSMENT FOR SOLVING THE PLASTIC WASTE CRISIS

CONVERSION OF WASTE PLASTIC INTO LIQUID FUELS

PLENARY SESSION: WASTE PLASTICS (INVITED TALKS)

POSTER SESSION: WASTE PLASTICS

RECYCLING AND UPCYCLING OF PLASTIC WASTE

566a The Role of the NIST Mass Spectral Library in the Identification of Plastic-Related Compounds: Extractables and Leachables

Adam Zuber, Yamil Simón-Manso, Edward P. Erisman, Tytus D. Mak, Meghan C. Burke Harris, William E. Wallace, Stephen E. Stein

Global consumption of plastic materials has thus far followed a primarily linear economic model, leading to an inordinately untenable situation from both environmental and economic forefronts. The National Institute of Standards and Technology (NIST) has recently launched the Circular Economy Program to address these concerns. In the vision of the circular economy, consumed materials are not merely discarded as waste but rather revalorized and reintroduced into the economy. NIST aspires to develop standard reference materials (SRMs), standard reference data (SRDs), and analytical techniques to serve that purpose.

In line with the Circular Economy Program, the NIST Mass Spectrometry Data Center (MSDC) has dedicated significant resources to developing SRDs for various plastic-related compounds (PRCs). To that end, efforts have been extended to identify PRC presence in human biofluids (e.g., blood, urine, breast milk), biomedical applications (e.g., infusion bags, drug containers), and drinking water, to name a few examples. In particular, the MSDC has developed a robust methodology to analyze extractables and leachables (E&L). This includes extending the existing NIST Mass Spectral Library (SRD 1A, NIST23) for PRCs and chemical analysis of E&L from standard commercial polymers (extracted using polar, semi-polar, and non-polar solvents) using LC-MS/MS – via Electrospray Ionization (ESI) and Atmospheric Pressure Chemical Ionization (APCI) techniques – and pyrolysis-GC-MS – via Electron Ionization (EI). Approximately one hundred of the most common polymers and commercially available mixtures have been analyzed in this way. The forthcoming release of the updated NIST Mass Spectral Library (NIST26) promises to showcase a vastly more comprehensive range of spectra for PRCs. Here, we review our efforts at the MSDC towards the NIST Circular Economy Program and NIST26 and consider potential avenues for future development.

566b Agricultural Plastic Waste: Analysis of Their Use, Quantification, and Management

Cesar Lubongo, Paschalis Alexandridis

Plastics are extensively used in agricultural activities, as they are used to transport feed and fertilizers, reduce development time for seed and fruit, cover crops, mitigate the growth of weeds, increase yields, increase water-use efficiency, retain moisture, wrap silage, and extend the lifetime and quality of fruit and produce. However, no adequate recycling infrastructure or market are in place for agricultural plastic waste (APW) which suffer from high contamination rates and degradation due to extended sunlight, pesticides, and soil exposure. Worse, APW release microplastics in the soil, and these microplastics contaminate our food and water.

In this study we set out to quantify the amount of plastic waste generated in New York State, identify its main constituents, and analyze management practices, with the aim to improve upon the current situation. Specifically, we quantify APW by collecting data through surveys and interviews with farmers, counties, and recycling companies, and by developing a prediction model of APW generation using a multiple regression analysis with the type of farm, revenue, and farm size as independent variables, and the quantity of APW generated as dependent variable. Drawing information from surveys and interviews, we identify challenges associated with management practices and recycling of APW.

Acknowledgements: The authors thank New York's Environmental Protection Fund, the New York State Department of Environmental Conservation (NYSDEC), for financial support.

566c Reuse of Plastic Packaging - When, Where and How Should We be Doing It?

Rachael Rothman, Maryam Hoseini, Stuart Walker, Alex Newman, Sarah Greenwood, Thomas Webb, Harriet Baird, Tony Ryan

Over 140 million tonnes of plastic packaging is produced annually around the world. The majority of this currently goes to waste, with a small amount being recycled, and around ⅓ leaves the managed waste system causing environmental harm and pollution. If some of this plastic packaging could be reused, the environmental impacts would be significantly less. Here we discuss the case studies of two major interdisciplinary projects: the UK based project Many Happy Returns that looks at take away containers, and the EU Horizon project Buddiepack, that investigates take away containers, refillable detergent bottles, trays for meat packing, catering trays for schools and "grab and go" take away. In each case, the break even point for global warming potential is calculated for the proposed reusable packaging - this indicates the number of times a reusable container must be used to be better than the single use equivalent. The impacts of packaging design are considered and the water usage evaluated. The break even points calculated are compared to the willingness of consumers to use worn containers.

Container lightweighting and use of renewable electricity are highlighted as key variables for environmental performance. The return rate of the container is particular important for the break even.

566d Economic and Environmental Investigation of Mechanochemical Technologies for Upcycling of Plastic Waste

Elisavet Anglou, Patricia Stathatou, Fani Boukouvala

The transition to circular economies stands as one of the most prevalent challenges of our times [1, 2]. Chemical recycling methods have emerged as a promising alternative as they allow the breakdown of polymers into their constituent monomers, enabling the material to be fully recycled. However, in contrast to conventional mechanical recycling, the complexity of chemical recycling processes have limited their implementation at the industrial scale [3]. One promising chemical recycling technology is mechanochemical depolymerization because of the high conversion rates achieved under mild conditions [2]. Our previous study indicated that the process can be economically viable for PET recycling at the industrial scale with identified economic metrics comparable to those conventional and alternative technologies [4, 5]. Nonetheless, integrating new technologies into the existing waste management network presents significant challenges because of the multiple stakeholders involved, various processing facilities, alternative options, and products [6].

To address these questions for mechanochemical recycling, this work applies a process system engineering framework that integrates value-chain, technoeconomic and life-cycle analyses. An equation-based formulation is adapted for this purpose, based on a realistic representation of the US recycling industry. The mathematical model is designed to minimize the environmental burden and the associated costs of the value-chain network, subject to constraints related to material balances and product quality requirements. Traditional waste management routes (e.g., landfilling, incineration, mechanical recycling) are incorporated within the superstructure network. Rigorous data from technoeconomic and life cycle analysis for mechanochemical depolymerization are detailed and incorporated within the modeling framework. Sensitivity analyses are performed to assess the impact of different design variables (e.g., efficiencies, economic parameters) to multiple objectives. Centralized as opposed to the existing decentralized market configuration scenarios are also examined. Further, the pareto-optimal solution is identified to discuss the trade-offs between environmental- and cost-friendly solutions for mechanochemistry. Finally, the results are compared with recent studies on competing recycling technologies [7, 8] to holistically assess the benefits and impacts of deploying mechanochemistry at the large scale. In summary, this study presents a computational framework that can be utilized to evaluate the employability of mechanochemical depolymerization, compare alternative pathways, and enable the design of cost-effective closed-loop value chains for waste plastics.

Citations

[1] I. Vollmer *et al.*, 'Beyond Mechanical Recycling: Giving New Life to Plastic Waste,' *Angew Chem Int Ed Engl,* vol. 59, no. 36, pp. 15402-15423, Sep 1 2020, doi: 10.1002/anie.201915651.

[2] A. W. Tricker *et al.*, 'Stages and kinetics of mechanochemical depolymerization of poly (ethylene terephthalate) with sodium hydroxide,' *ACS Sustainable Chemistry & Engineering,* vol. 10, no. 34, pp. 11338-11347, 2022.

[3] U. S. Chaudhari *et al.*, 'Systems analysis approach to polyethylene terephthalate and olefin plastics supply chains in the circular economy: A review of data sets and models,' *ACS Sustainable Chemistry & Engineering,* vol. 9, no. 22, pp. 7403-7421, 2021.

[4] E. Anglou, Y. Chang, A. Ganesan, S. Nair, C. Sievers, and F. Boukouvala, 'Discrete Element Simulation and Economics of Mechanochemical Grinding of Plastic Waste at an Industrial Scale,' in *Computer Aided Chemical Engineering*, vol. 52: Elsevier, 2023, pp. 2405-2410.

[5] E. Anglou *et al.*, 'Process development and techno-economic analysis for mechanochemical recycling of poly (ethylene terephthalate),' *Chemical Engineering Journal,* p. 148278, 2023.

[6] J. Ma *et al.*, 'Economic Evaluation of Infrastructures for Thermochemical Upcycling of Post-Consumer Plastic Waste,' 2022.

[7] T. Uekert *et al.*, 'Technical, economic, and environmental comparison of closed-loop recycling technologies for common plastics,' *ACS Sustainable Chemistry & Engineering,* vol. 11, no. 3, pp. 965-978, 2023.

[8] Y. Luo, E. Selvam, D. G. Vlachos, and M. Ierapetritou, 'Economic and Environmental Benefits of Modular Microwave-Assisted Polyethylene Terephthalate Depolymerization,' *ACS Sustainable Chemistry & Engineering,* vol. 11, no. 10, pp. 4209-4218, 2023.

566e A Quantitative Circular Economy Assessment Framework for Plastic Waste Management Systems

Paola Alejandra Munoz Briones, Aurora del Carmen Munguía-López, Horacio Aguirre-Villegas, Styliani Avraamidou

Plastic production and use have increased nearly 200-fold since their first appearance due to their versatility and outstanding benefits.[1] However, plastic production leads to plastic waste generation that pollutes the environment and threatens the well-being of wildlife and humans. In 2018, nearly 13.6% of plastic containers and packaging generated were recycled, only 16.9% were combusted for energy, and over 69% ended up in landfills.[2] Plastic packaging remains as one of the most important challenges given its immediate disposal after first use. Circular Economy (CE), as an alternative system, promotes the transition towards sustainability by targeting actions to minimize environmental pollution and promote the closed-loop handling of resources, preserving the value of materials and products for an extended period. Therefore, identifying the optimal waste management technologies under the circularity concept will accelerate the transition to sustainability. Circular Economy assessment frameworks, propose a complete assessment of environmental impacts, expanding beyond greenhouse gas emissions and involving impacts like resource depletion and the reduction of waste simultaneously. To effectively evaluate and compare the circularity of different processes, a quantitative metric is necessary. Some studies suggest CE indicators, but they often concentrate only on a subset of the CE goals or present a narrow scope applicable just to specific products.[3]

This work aims to adapt a CE framework previously proposed to be applicable for assessing the circularity of different plastic waste management technologies. Our previous work considers the CE main goals and characteristics and proposes indicators to calculate circularity subindexes for specific categories and an overall circularity index.[4] This work develops a new substitutability category that through the corresponding index will be able to represent the capacity of recycled plastics to replace virgin plastics.[5,6] Besides considering waste, emissions, water, procurement, and energy, the substitutability of plastics is crucial to demonstrate how effectively plastics recycling can reduce the production of plastics and their associated environmental implications. For this category, factors such as mechanical and processing properties, color, and smell define the corresponding indicators and metrics. The recycled plastic properties are compared to reference values of virgin plastic properties. By incorporating the substitutability criteria and applying the circularity metric to different waste management technologies, we expect to identify which technologies are more environmentally friendly and isolate possible improvement areas of the technology in different categories like waste, emissions, or substitutability. To demonstrate the applicability of the framework, the circularity of mechanical recycling, pyrolysis, solvent-targeted recovery, and precipitation (STRAP^{TM}),[7] landfilling, and incineration is evaluated and presented.

References

1. Wang, C., Liu, Y., Chen, W. Q., Zhu, B., Qu, S., & Xu, M. (2021). Critical review of global plastics stock and flow data. Journal of Industrial Ecology, 25(5), 1300-1317.
2. EPA. (2023). Containers and Packaging: Product-Specific Data. https://www.epa.gov/facts-and-figures-about-materials-waste-and-recycling/containers-and-packaging-product-specific
3. Kristensen, H. S., & Mosgaard, M. A. (2020). A review of micro level indicators for a circular economy–moving away from the three dimensions of sustainability?. Journal of Cleaner Production, 243, 118531.

4. Baratsas, S. G., Pistikopoulos, E. N., & Avraamidou, S. (2022). A quantitative and holistic circular economy assessment framework at the micro level. Computers & Chemical Engineering, 160, 107697.
5. Chialdikas, E., Munguia-López, A. D. C., Aguirre-Villegas, H., & Avraamidou, S. (2023). A framework for the evaluation of the circularity of plastic waste management systems: a case study on mechanical recycling of HDPE. Foundations of computer aided process operations/chemical process control: In-Press.
6. Demets, R., Van Kets, K., Huysveld, S., Dewulf, J., De Meester, S., & Ragaert, K. (2021). Addressing the complex challenge of understanding and quantifying substitutability for recycled plastics. Resources, Conservation and Recycling, 174, 105826.
7. Walker, T.W., Frelka, N., Shen, Z., Chew, A.K., Banick, J., Grey, S., Kim, M.S., Dumesic, J.A., Van Lehn, R.C. and Huber, G.W. (2020). Recycling of multilayer plastic packaging materials by solvent-targeted recovery and precipitation. Science advances, 6(47)

566f Grassroots Innovation in Plastic Waste Management: Ishara's Model.

Akorfa Dagadu

Ishara is pioneering a grassroots initiative in Ghana to address the escalating plastic waste crisis through innovative technology and community engagement. Our mission is to create a closed-loop system that empowers local waste pickers, fosters sustainable recycling practices, and nurtures a culture of environmental responsibility.

At the heart of Ishara's approach is a mobile app designed to streamline the plastic waste collection process, making it more efficient and transparent. This app connects individual users and waste pickers, enabling them to easily schedule and track waste pickups. It also directly links to recycling companies, ensuring a consistent supply of segregated and cleaner recyclables.

Through our collaboration with the University of Ghana, we have initiated primary market research to understand and integrate local recycling habits, challenges, and opportunities into our model. Starting with such universities, we are on track to redirect over 25 tons of plastic waste daily, plastics that would have otherwise polluted our environment. Our initiative aims to mitigate environmental pollution and offer fair economic opportunities for waste pickers, often marginalized in the waste management ecosystem.

In presenting at AIChE, we seek to explore innovative solutions and collaborations to enhance our system's efficiency and impact. We believe that addressing the plastic waste crisis requires a multifaceted approach, combining technological innovation with community-based strategies to achieve sustainable change and economic empowerment.

Energy Consumption - Carbon Emissions Analysis and Technology Layout Optimization For Plastics Industry At The Life Cycle: Country-Based Study

Menglin Liu, Zhikai Lang, Xianyang Meng and Jiangtao Wu, MOE Key Laboratory of Thermo-Fluid Science and Engineering, Xi'an Jiaotong University, Xi'an, Shaanxi, China

Paper presentation ID# 693812

Abstract

The plastics industry is a significant consumer of fossil raw materials. With the continuous expansion of future plastics production scale, it becomes crucial to reasonably and accurately predict the future development trend of plastics production. To address the high energy consumption and carbon emission issues throughout the entire life cycle process of the plastics industry, multidimensional interventions are imperative across all stages. These interventions aim to optimize the technological layout and formulate development plans for energy conservation and emission reduction in the future plastics industry. This work is centered on predicting the developmental trajectory of plastics production and conducting simulations alongside multi-objective optimization to enhance energy efficiency and reduce emissions in the plastics industry. The CRITIC combination model of plastics production and the LEAP model based on the results of LCA analysis are constructed. Through scenario simulations, full life cycle cost-effectiveness models and multi-objective optimization frameworks are developed for analyzing the economic viability of energy-saving and emission reduction initiatives in the future plastics industry. These models further aid in optimizing the technological landscape of the plastics industry under various objectives. The application of this methodology to China's plastics sector enables the forecasting of the country's total plastics production, the energy consumed during the entire life-cycle of these products, and the carbon emissions generated up to 2030. Additionally, the potential of different technologies for energy conservation and emission reduction are assessed. CCUS technology and clean energy substitution are necessary ways to reduce emissions, with the net investment costs of 38 and 19 dollar/tCO_{2e}, respectively. Improving the utilization rate of plastic waste meets the requirements of industry development with low energy consumption and carbon emissions, but its initial cost will be up to 134 dollar/tCO_{2e}. In 2030, the net investment demand of the plastic industry will be 343 million dollars. The analysis presents a number of alternative transformations within the plastic industry, offering those responsible for decision-making in private enterprises and utilities the greatest possible flexibility in decision-making processes. Concurrently, this methodology encapsulates the fundamental principles of creative innovation, which are essential for effectively addressing significant challenges.

1. Introduction

With the vigorous development of the chemical industry, plastic products have been widely used in all walks of life. Since 1950, the world's annual plastics production has grown exponentially[1], from 2 million tons to 460 million tons in 2019[2]. As the production, transportation, consumption and post-treatment of plastic products are characterized by high energy consumption, carbon emissions and pollution, the rapid expansion of the plastics industry has caused a series of energy, climate and ecological problems.

China is the largest plastic producer and consumer in the world[3], so it is more urgent than most other countries in "plastic control"—regulating plastic production, consumption and post-treatment. By 2019, China's plastics production in primary form had accumulated to 1 billion tons[3]. The research by the Climate Change and Energy Transformation Project of Peking University shows that the plastics industry has become the second largest oil consumption field in China after transportation, and it is estimated that by 2050, the CO_2 generated by plastics will reach about 1.3-1.4 billion tons, accounting for almost all the remaining carbon budget under China's carbon neutral target. Furthermore, China's plastic waste output reached 63 million tons in 2019[4]. It will take hundreds or even thousands of years for these plastics to degrade in the natural environment, which poses a serious burden on the ecosystem. In response to this series of problems, China implemented the "Plastic Restriction Order" in 2008[5], and in recent years has issued the stricter "Plastic Prohibition Order"[6]. However, the current intervention measures are mainly focused on the plastics production and waste disposal, the governance system is not complete and the measures are not detailed. With the aggravation of energy shortage and the proposal of the "3060" carbon target, it is necessary to propose stricter and more detailed policy measures and quantitative indicators for each process of the plastics industry's life cycle.

This study uses the LCA method to quantitatively analyze the energy consumption and CO_2 emissions of plastics in China, and proposes multi-dimensional interventions from the source of production to the end of disposal. Using the results of the life cycle data as the input parameters, the LEAP-Plastic model is established to forecast the future energy consumption and carbon emissions of China plastics industry under different scenarios. Then, a cost-benefit model of energy saving and emission reduction in the life cycle of the plastics industry is established to predict and evaluate the feasibility and commercial value of various technologies from an economic perspective. Through the analysis, prediction and cost-benefit accounting, it is expected to provide a reference for relevant government departments to propose more comprehensive and enforceable action plans for energy conservation, emission reduction and pollution control in the plastics industry.

2. Methodology

2.1 LCA model

This study established a LCA model for China plastics industry, defined its research framework based on GB/T24040, and analyzed the energy consumption and carbon emissions of the life cycle chain[7]. The life cycle of the plastics industry includes four stages: production, transportation, consumption and post-treatment. In terms of production, in addition to five general-purpose resins, the production and processing of plastic additives is also included. In respect of transportation, the plastic transportation is mainly highway and railway transportation. Apart from plastic products, the transportation of raw materials and additives is also involved. In the matter of post-treatment, its methods mainly contain abandonment, landfill, incineration and recycling. The research framework and system boundary description of this LCA model are shown in Fig. 1.

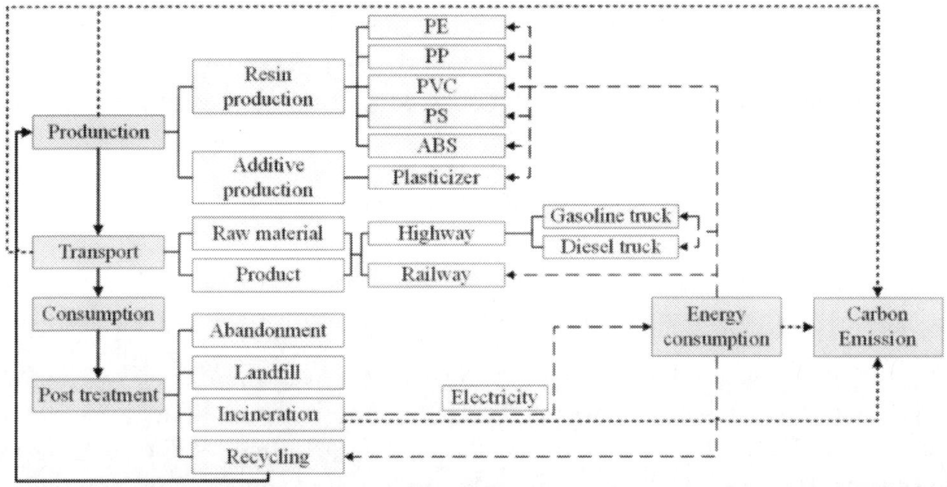

Fig. 1. Research framework and system boundary of the LCA model

2.2 LEAP-Plastic model

The LEAP-Plastic model was established, which includes four modules: terminal energy demand, energy processing and conversion, energy supply and greenhouse gas emissions. Without considering the import and export of energy and products, the time span of this model is 1990~2021, the base year is 2021, and the prediction period is 2022~2030. The structural framework of this model is shown in Fig. 2. Compared with other LEAP prediction models, this model is based on the LCA data results as the input parameters, the quoted data can more accurately reflect the actual situation of China plastics industry, and the calculation results effectively cover the life cycle processes of the plastics industry.

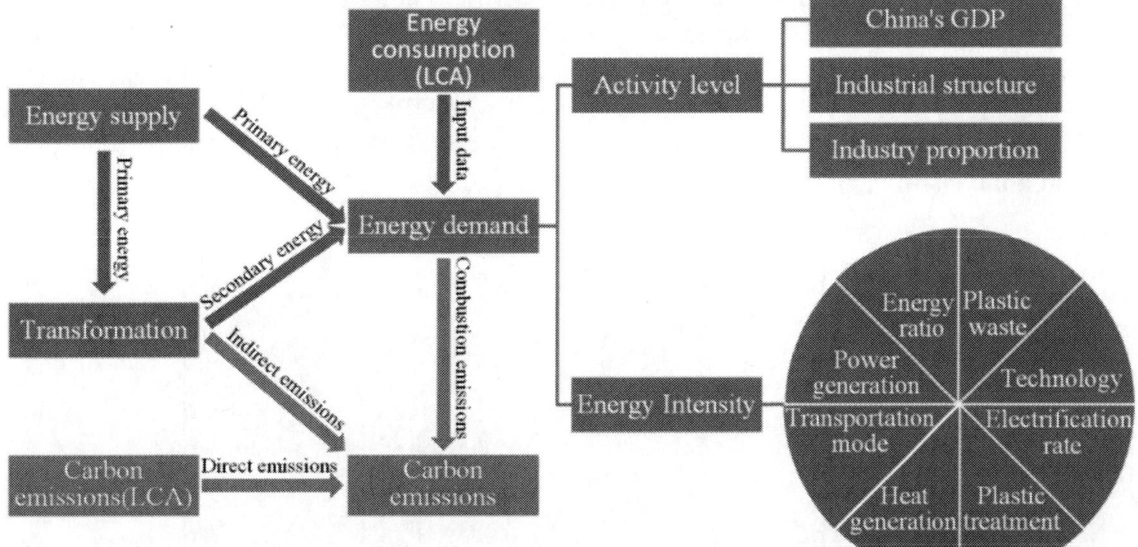

Fig. 2. LEAP-Plastic model framework

2.3 Cost-benefit model of energy saving and emission reduction

Based on energy economics, a cost-benefit model of energy saving and emission reduction for the life cycle of China plastics industry was established, investment costs and economic benefits generated from the BAS to CER scenario in 2030 were analyzed and forecasted. The investment costs usually consider the technology investment costs and operation costs. The technology investment costs include the equipment purchase costs, project construction costs, equipment installation costs, etc. The operation costs include materials costs, labor costs, management costs, maintenance costs, depreciation costs[8]. Moreover, it's assumed that the energy-saving technology can achieve the purpose only by saving the primary energy and purchased electricity and heat. Meanwhile, as there is no air pollutant trading market in China, the emission reduction benefit only considers CO_2 and are exclusive of other conventional air pollutants. Mathematically, the cost-benefit model is summarized as follows:

$$P_{net} = P_i - P_e - P_c \tag{1}$$

$$P_i = \sum_b \left(Q_{bt} \times P_{b,it} + P_{b,iL} \div L_{b,u}\right) \tag{2}$$

$$P_e = \sum_k \left(E_{k,tr} \times P_k\right) - \sum_v \left(E_{v,ti} \times P_v\right) \tag{3}$$

$$P_c = C_r \times P_{ce} \tag{4}$$

where P_{net} is the net investment added value. P_i refers to the investment costs. P_e and P_c represent energy saving and emission reduction benefits, respectively. Q_{bt} implies the total usage scale of the b technology. $P_{b,it}$ denotes the unit annual operating costs. $P_{b,iL}$ indicates the technology investment costs. $L_{b,u}$ represents the service life. $E_{k,tr}$ and $E_{v,ti}$ are the reduction of the k energy consumption and the increase of the v energy consumption, respectively. P_k and P_v refer to the unit prices of the k and v energy sources, respectively. C_r implies the total emission reduction, P_{ce} denotes the unit price of carbon trading.

3. Data collection and scenario setting

3.1 LCA model

The data related to the energy consumption of the LCA model in China plastics industry are sorted out. The annual output of plastics is taken from China Statistical Yearbook (1991-2021)[9], and the results are shown in Fig. 3. The output of five general-purpose plastics (PE, PP, PVC, PS and ABS) accounts for 24%, 28%, 36%, 6% and 6%, respectively[10]. What's more, PE are divided into linear low density polyethylene (LLDPE), low density polyethylene (LDPE) and high density polyethylene (HDPE), of which LLDPE and HDPE account for about 96% of the total production capacity in 2020. Therefore, this study doesn't consider the production of LDPE and assumes that LLDPE and HDPE account for 55% and 45% of the PE produced, respectively[11]. Then, the data required for energy consumption analysis in each sub process is collected.

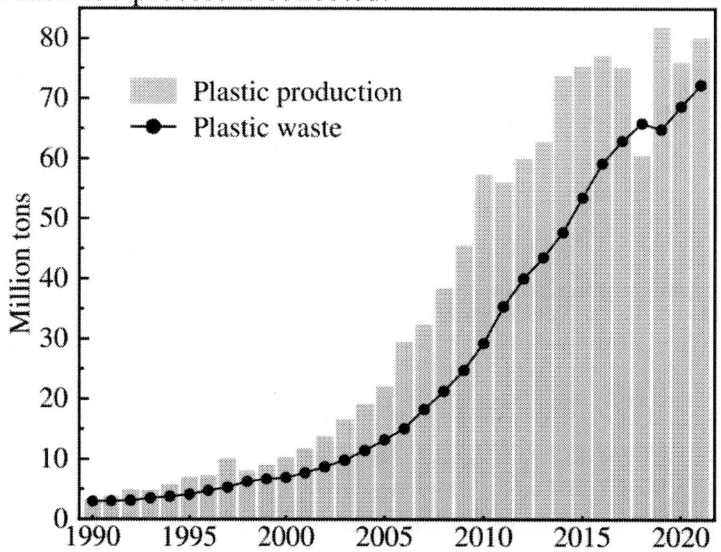

Fig. 3. Annual production of plastics and waste in China

The direct emissions from the plastics industry originate from many chemical processes involved in the production. Through the LCA of plastics, it is concluded that the direct emissions mainly include two sources. First, when PVC is produced by the ethylene method, a large amount of waste gas generated by the reaction, the direct carbon emissions generated by each ton of PVC production are 345 $kgCO_{2e}$[12]. In addition, CO and CO_2 are also directly generated through chemical reaction during the production of acrylonitrile. And it's calculated that the direct carbon emissions of producing one ton of acrylonitrile are 93.34 $kgCO_{2e}$[13]. The potential carbon emissions of purchased electricity and heat need to be specifically analyzed according to the current situation of China's power and heat industry. The power generation mode in China is mainly coal-fired power generation, followed by hydroelectric, wind, nuclear, gas, solar, biomass and oil-fired power generation.

3.2 Scenarios

In order to analyze the impact of various policies and technologies on energy conservation and emission reduction, this study established several scenarios based on the development forecast of social, economic, technological and policy factors. The baseline scenario (BAS) is a continuation of the current policies, technologies and social development, and some scholars have predicted the development trends of China's GDP, industrial structure, energy intensity and other parameters under this scenario, which provides a certain basis and reference for this study. Meanwhile, considering the main factors affecting the future energy demand and CO_2 emissions, five scenarios are set based on the BAS scenario, and the specific settings are shown in Table 1.

Table 1. Parameter settings for five scenarios

Scenario	Abbreviations	Parameter setting
Clean Energy Generation Scenario	CEG	Improve the application of new energy and natural gas in power and heat generation.
Energy Efficiency Improvement Scenario	EEI	Increasing the rate of energy intensity reduction in the production and post-treatment processes from 1.25% per year in the BAS scenario to 1.5%, and the transportation process from 1.5% to 2%[14].
CCUS Technology Scenario	CCUS	The emission reduction efficiency of CCUS is 90%[15].
High Recycling Rate Scenario	HRR	This scenario significantly improves the recycling rate of plastic waste, and increase the power generation rate of plastics incineration.
Comprehensive Emission Reduction Scenario	CER	Combining the above four technologies and policies for energy conservation and emission reduction, a CER scenario is proposed.

4. Results and discussion

4.1 The LCA of China plastics industry

The life cycle energy consumption and carbon emissions of five general-purpose plastics are analyzed, and the results are shown in Fig. 4. Through analysis, it can be concluded that the average life cycle energy consumption of five general-purpose plastics in China is 2.03 tce/t, of which the maximum is 2.37 tce/t for ABS plastics and the minimum is 1.86 tce/t for PVC plastics. The energy consumption of ABS plastics is significantly higher than that of other plastics, mainly due to the complexity of its production process and the high consumption of raw materials.

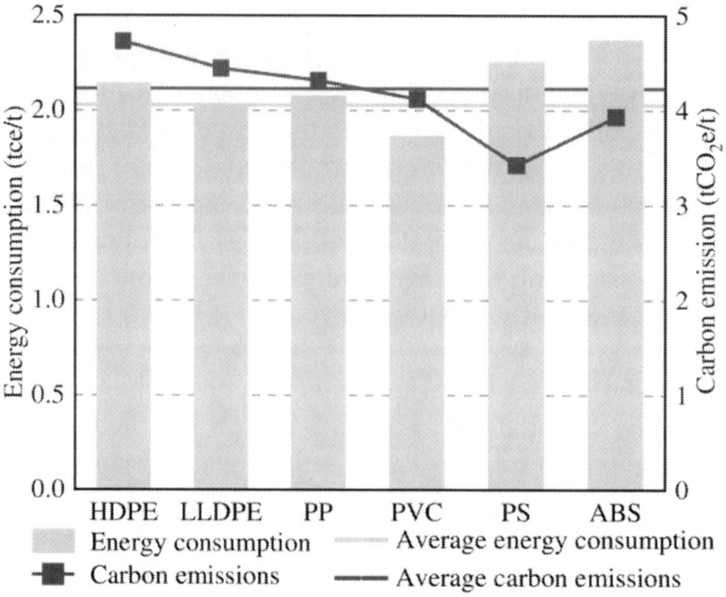

Fig. 4. The energy consumption and carbon emissions of five general-purpose plastics

The life cycle energy consumption of China plastics industry from 1990 to 2021 is analyzed, and the results are shown in Fig. 5. The overall energy consumption shows a trend of rising first and then fluctuating, which is highly consistent with the change trend of plastic output. In 2018, with the decline of plastics production, the energy consumption and carbon emissions decreased significantly. The reasons are as follows: The Sino-US trade war led to the reduction of OEM and slow export. The Chinese government banned the entry of foreign waste[16], which led to a sharp decline in the output of the recycled plastics industry, and according to the customs statistics, 51800 tons of recycled plastics waste were imported in 2018, down 99.11% from 2017. Domestic environmental protection is strict, and small and micro enterprises are phased out. In addition, according to the data released by the National Bureau of Statistics of China, the total energy consumption of China in 2021 was 5.24 billion tce[17], while the plastics industry consumed 160 million tce, accounting for 3.1% of total energy consumption.

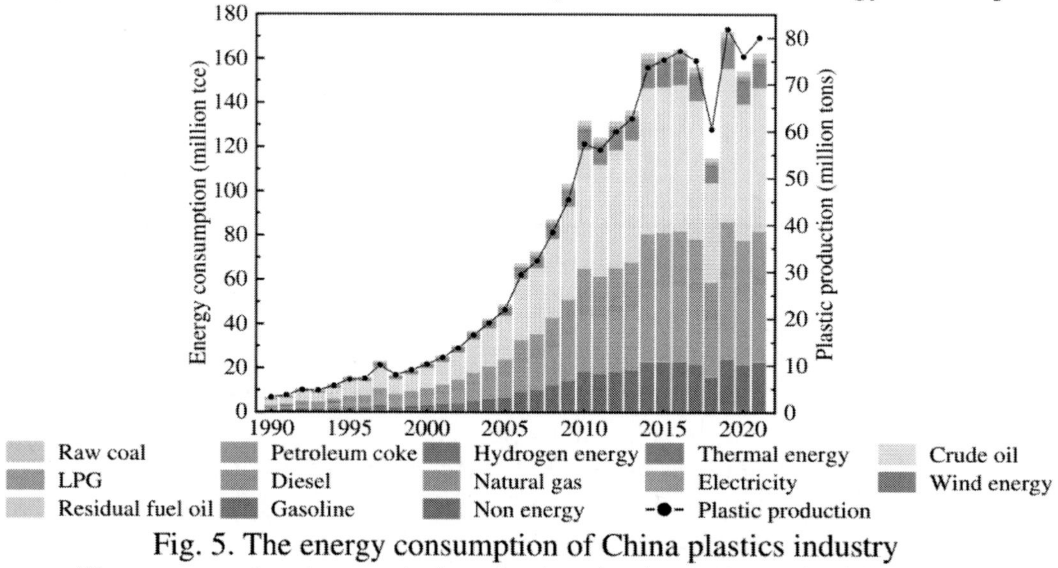

Fig. 5. The energy consumption of China plastics industry

The specific sources of carbon emissions in the plastics industry in 2021 are analyzed, and the proportion is shown in Fig. 6. The indirect carbon emissions of purchased electricity account for 64.7%, so the clean and low-carbon development of the electricity industry has a significant impact on the emission reduction of the plastics industry. Plastic waste incineration is the second largest source, accounting for 12.5%. As the proportion of plastic waste incineration increases, clean incineration technology should be vigorously developed and CCUS should be equipped to reduce environmental pollution. The proportion of carbon emissions from fossil energy incineration is 19.9%, of which LPG accounts for the highest proportion of 11%. The incineration of fossil energy is a key area that is difficult to reduce emissions at present. It is necessary to reduce the actual carbon emissions by improving energy efficiency and pay attention to carbon removal technologies such as CCUS to achieve deep decarburization. In addition, the direct carbon emissions of production and the potential carbon emissions of purchased heat account for 2.6% and 0.4%, respectively.

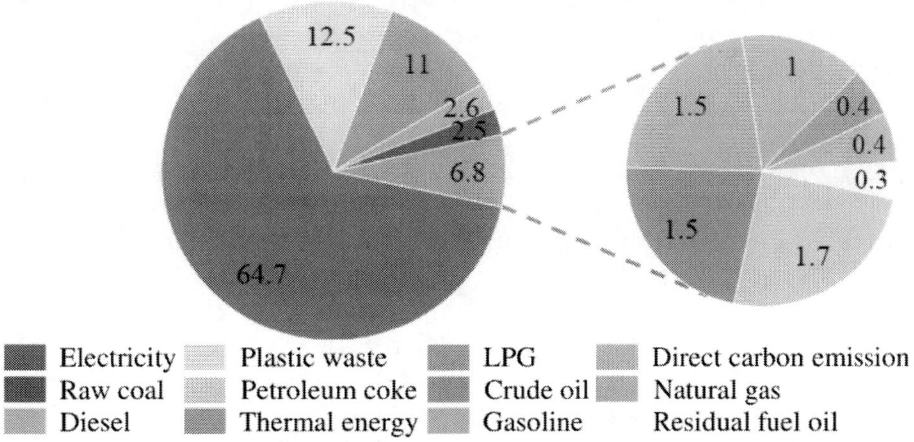

Fig. 6. The proportion of various carbon emissions

4.2 Development forecast of China plastics industry

The specific consumption of each energy and the energy-saving effect of each technology under different scenarios in 2030 are analyzed, as shown in Fig. 7. Improving energy efficiency can effectively reduce the actual use of various energy sources, especially electricity, as the main form of energy supply, is most affected. Increasing the recycling rate of plastic waste can significantly reduce the consumption of crude oil and non-energy raw materials, and also reduce the energy consumption in the production process. Compared with the BAS scenario, the CER scenario will save 14.48 million tce in 2030, of which 28% will be saved by improving energy efficiency, and 72% will be saved by heightening plastic recycling and incineration rates. However, the energy saving of relevant technologies and policies only accounts for 8.2% of the total energy consumption in the CER scenario, which requires more stringent industrial energy-saving policies and faster popularization rates of advanced technologies and equipment.

The prediction of carbon emissions in China plastics industry throughout the life cycle is shown in Fig. 8. The forecast results of the BAS scenario and four sub-scenarios by 2030 are all on the rise, while the CER scenario shows a downward trend year by year. Under the BAS scenario, the carbon emissions in 2030 are expected to reach 400 million tCO_{2e}, an increase of about 20 million tCO2e from 2021. Compared with the BAS scenario, the CER scenario reduces a total of 213 million tCO2e from 2022 to 2030. The emission reduction effect under the existing policies lags behind the "3060" carbon target. From the forecast results of the four sub-scenarios, it can be seen that improving plastic recycling and incineration rate is the most significant measure to reduce emissions in the short term, and CCUS is the technology with the most potential in the long term. However, the implementation of any single policy or technology will not enable the plastic industry to achieve the carbon target of peaking in 2030, which requires the synergy of multiple technologies and policies.

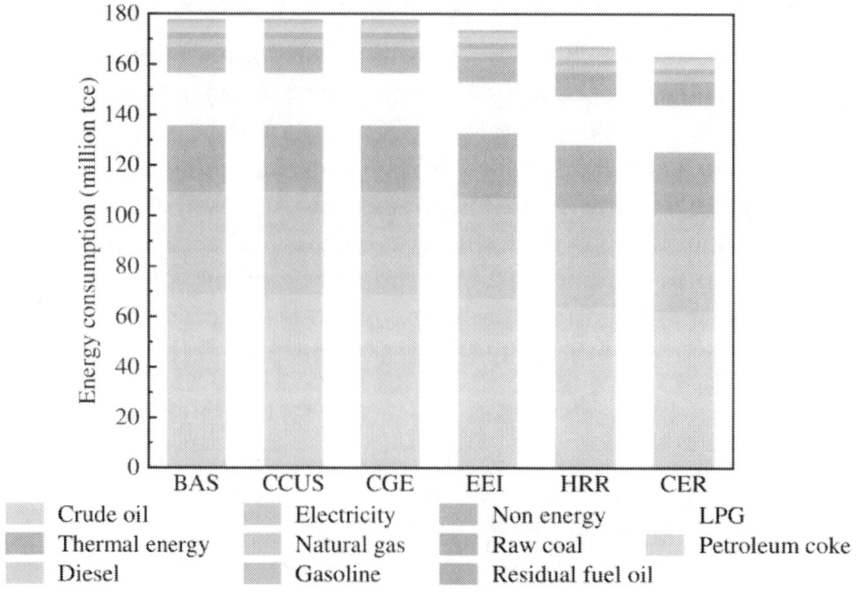

Fig. 7. Energy consumption under different scenarios in 2030

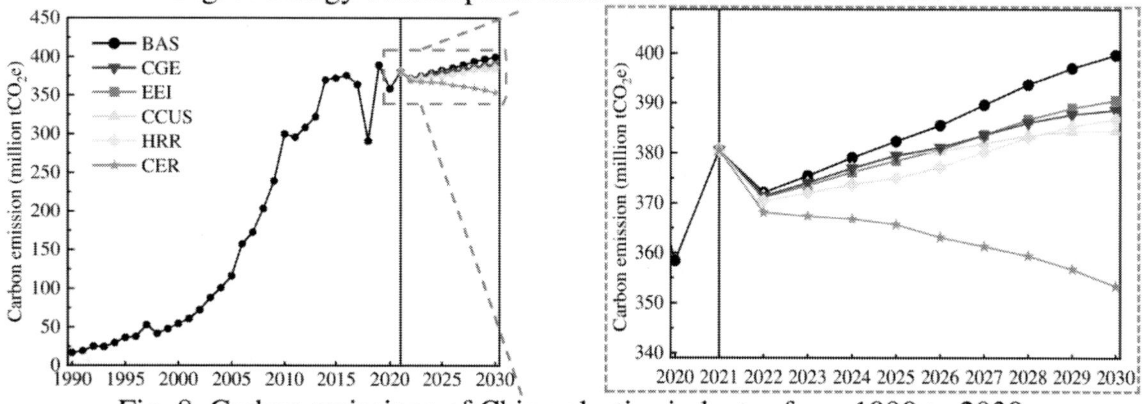

Fig. 8. Carbon emissions of China plastics industry from 1990 to 2030

4.3 Cost-benefit analysis of energy conservation and emission reduction

The investment costs and benefits of energy conservation and emission reduction in the plastics industry in 2030 are analyzed, and the results are shown in Fig. 9. Compared with the BAS scenario, the total investment costs increment of the CER scenario is about 60.5 billion yuan, of which the proportion of HRR sub-scenario is as high as 77%. The reason is that the recycling costs of waste plastics and the purchase costs of material during plastic incineration are relatively high. Of course, the HRR scenario will also create a lot of income for enterprises due to the production of recycled plastic products. The emission reduction and energy conservation benefit of the CER scenario are 10 and 32.6 billion yuan respectively, which is not enough to offset the investment costs, and requires an additional net investment of 17.8 billion yuan. Nevertheless, without considering the recycling costs of plastic products, the total investment cost is only 22.9 billion yuan, indicating that the policies and technologies for energy saving and emission reduction will create economic value on the whole by 2030.

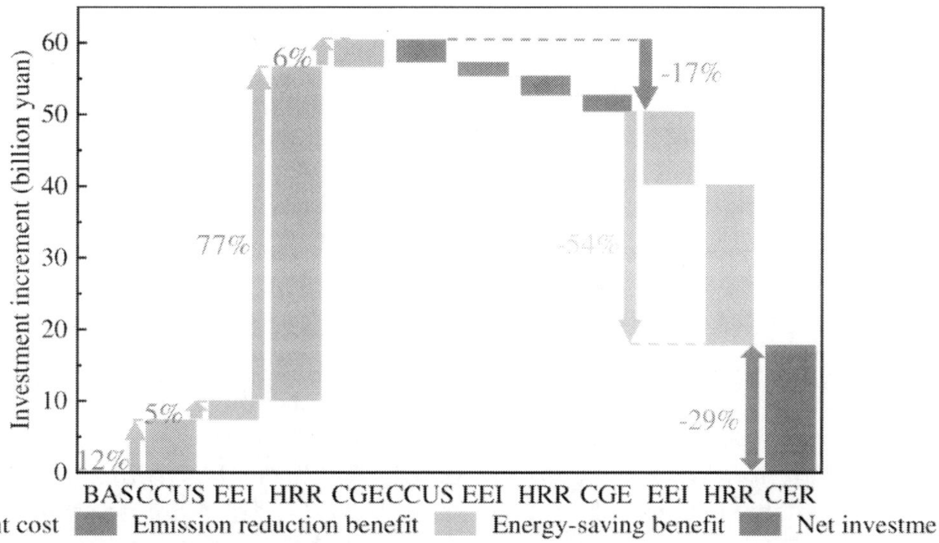

Fig. 9. Cost-benefit analysis of energy saving and emission reduction

5. Conclusion

In this study, a LEAP-Plastic model based on the LCA is established to predict the future development trends of China plastics industry under different scenarios by 2030, and the energy conservation and emission reduction potential of various technologies are further discussed. Ultimately, a cost-benefit model is built to analyze the investment costs and energy conservation and emission reduction benefits. The specific conclusions are as follows:

(1) The average energy consumption and carbon emissions of five general-purpose plastics are 2.03tce/t and 4.23tCO2e/t. Meanwhile, the energy consumption and carbon emissions of the plastics industry in 2021 account for 3.1% and 3.2% of China's total, respectively. Crude oil, as the main fossil raw material for plastics production, accounts for more than 40% of the energy consumption. Electricity accounts for more than 20% as the main form of energy supply. In 2021, electric consumption and plastic incineration are the main sources of carbon emissions, accounting for 64.7% and 12.5%.

(2) The energy consumption and carbon emissions of China plastics industry from 2022 to 2030 are forecasted. Under the BAS scenario, the energy consumption and carbon emissions in 2030 will reach 178 million tce and 400 million tCO2e. Compared with the BAS scenario, the cumulative energy saving and emission reduction of the CER scenario by 2030 will be 74.12 million tce and 213 million tCO2e. In 2030, the two sub-scenarios of EEI and CER contribute 28% and 72% of energy conservation, respectively. The highest contribution rate of emission reduction is 32% of the CCUS scenario, and the lowest is 19% of the EEI scenario.

(3) The investment costs and benefits of the plastics industry are analyzed based on the LCA and LEAP-Plastic model. By comparison with the BAS scenario, the investment increment of the CER scenario in 2030 is 60.5 billion yuan, and the benefits of energy conservation and emission reduction are 10 billion yuan and 32.6 billion yuan, respectively. It's researched that improving the energy efficiency of the plastics industry will be the primary choice for the future development. CCUS technology and clean energy substitution are the essential ways to achieve the deep decarburization of the plastics industry, but they will not have commercial application value in 2030, which requires the support of national policies and rapid development of low-cost technologies. Improving the recycling rate of waste plastics is a necessary trend for the sustainable development, but a large amount of investment in the early stage needs the support of government policies and the financing and transformation of enterprises.

References

1. Gandhi, N., Farfaras, N., Wang, N. H. L., Chen, W. T., (2021). Life cycle assessment of recycling high-density polyethylene plastic waste. *Journal of Renewable Materials* 9, 1463-1483.
2. OECD (2022), Global plastics outlook: Economic drives, environmental impacts and policy options. OECD, Paris. https://www.oecd-ilibrary.org/content/publication/de747aef-en
3. Cao, S. Y., Chen, B. S., Guo, H., Sun, J. L., (2020). Plastics environmental footprint in China. BIPT, Beijing. http://www.nrdc.cn/Public/uploads/2020-12-21/5fe005fdf3d22.pdf
4. Sun, K., (2021). Study on aromatics production from catalytic pyrolysis of waste plastics. Doctor. Zhejiang University.
5. GOSCPRC (2008). Notice of the general office of the state council on restricting the production and sales of plastic shopping bags. GOSCPRC, Beijing. http://www.gov.cn/zhuanti/2015-06/13/content_2879030.htm
6. Liu, J., Fu, R., Wang, L., Sun, Y., Huang, T., (2021). Determination of organic carbon in biodegradable materials by total organic carbon solid sampling system. *Huanjing Huaxue-Environmental Chemistry* 40, 2268-2270.
7. Chen, L., Liu, M., Huang, J., (2009). GB/T 24040-2008 Environmental Management-Life Cycle Assessment-Principles and Framework national standard understanding. *Standard Science* 46, 76-80.
8. Gong, Y., (2022). Study on emissions and cost of reduction of multi-pollutants and CO_2 from cement industry in China. Master. Zhejiang University.
9. NBSPRC (2022). China Statistical Yearbook. China Statistics Press, Beijing. https://data.cnki.net/yearbook/Single/N2021110004
10. Kong, L. Q., (2016). The top priority of the plastics industry is to control the speed of capacity expansion. *China Rubber/Plastics Technology and Equipment* 42, 99-100.
11. Liu, C. Y., (2021). Impace of domestic polyethylene product structure change on α-olefins. *Petroleum & Petrochemical Today* 29, 16-21.
12. Xuan, S. Q., (2021). Treatment of waste gas from incineration in vinyl chloride production by ethylene process. *China Chlor-Alkali* 41, 20-21+41.
13. Lu, H. W., (2014). Analysis and optimization on critical factors in industrial operation of acrylonitrile unit. Master. Hefei University of Technology.
14. CNPCETRI (2020). Energy outlook 2050. CNPCETRI, Beijing. https://pdf.dfcfw.com/pdf/H3_AP202012251443823759_1.pdf?1608923532000.pdf
15. Cai, B. F., Li, Q., Zhang, X., (2021). China carbon dioxide capture, utilization and storage (CCUS) annual report (2021). Ministry of Ecology and Environment of the People's Republic of China, Beijing. https://img76.hbzhan.com/4/20210727/637629742157746067210.pdf
16. GOSCPRC (2017). Implementation plan for promoting the reform of solid waste import management system by prohibiting the entry of foreign waste. GOSCPRC, Beijing. http://www.gov.cn/xinwen/2017-07/27/content_5213919.htm
17. NBSPRC (2022). Statistical bulletin of the people's Republic of China on national economic and social development in 2021. NBSPRC, Beijing. http://www.gov.cn/xinwen/2022-02/28/content_5676015.htm

505a A Microkinetic Model for Isomerizing Ethenolysis

Ziqiu Chen, Aaron Sadow, Baron Peters

In isomerizing ethenolysis, tandem double-bond isomerization and olefin metathesis catalysts convert polyethylene and ethylene co-reactants into propylene. Isomerizing ethenolysis is particularly interesting among polymer upcycling strategies because of its potentially high selectivity to a specific value-added product. Following a theoretical analysis by Guironnet and Peters [*J. Phys. Chem.* 124, 3935 (2020)], Conk et al. [*Science*, 377, 1561 (2022)] demonstrated isomerizing ethenolysis in experiments using an iridium pincer dehydrogenation catalyst, a dimeric Pd(I) bromide isomerization catalyst, and a second-generation Hoveyda-Grubbs metathesis catalyst. This paper compares model predictions to the two-stage dehydrogenation and isomerizing ethenolysis experiments of Conk et al. In a model that accounts for the initial dehydrogenation and subsequent evolution of the chain length distribution, we show that the experimental propylene generation rates are consistent with an isomerizing ethenolysis rate that is zeroth order in the concentration of long chain ends. In contrast, Guironnet and Peters assumed a first order dependence on chain ends. To understand the discrepancy, we develop and solve a microkinetic model for the isomerizing ethenolysis reaction. Rate parameters in the microkinetic model are estimated from prior experiments and known equilibria. We find that the experiments of Conk et al. are performed near conditions of the theoretical maximum propylene production rate, where the kinetics are saturated with respect to both chain end and ethylene concentrations. For long chains, not pre-shortened by initial dehydrogenation and ethenolysis steps as in Conk et al., the model predicts lowered chain end concentrations and smaller propylene production rates that can become inhibited by high ethylene pressure.

505b A Integrated Mechanism and Mathematical Modeling for Suppressing Gas Formation in Ru-Catalyzed Hydrogenolysis

Yae Chan Park, Kyeongsu Kim, Chun-Jae Yoo, Ung Lee

The accumulation of plastic waste has led to a severe ecological crisis, exacerbated by inadequate waste management systems. To address plastic pollution, research on catalytic depolymerization has gained traction as a method to convert polyethylene into value-added materials, such as naphtha. However, limited research exists on the kinetic modeling of this process, which could reveal critical insights into its chemical and physical dynamics. This study proposes a stochastic mathematical model for the Ru-catalyzed depolymerization of HDPE, specifically via hydrogenolysis, where the metal-based mechanism facilitates C–C bond cleavage through alkane hydrogenation at a monofunctional metal site. Ru-based mechanisms have demonstrated potential for depolymerizing under mild conditions, though excessive depolymerization may produce gaseous byproducts, such as CH_4. The aim of this study is to establish a low-dimensional stochastic model for the hydrogenolysis process, utilizing kMC simulation to predict experimental outcomes accurately and optimize conditions. This research ultimately seeks to inform reactor design and optimization, focusing on parameters such as temperature, H_2 pressure, and Ru particle size.

505c Hydrothermal Liquefaction of Post-Consumer Recycled Linear Low-Density Polyethylene with Different Molar Mass in Supercritical Water

Kalsoom Jan, Taofeng Lu, Ana Paula Azeredo, Regina Nonemacher, Raquel S. Mauler, Grace Chen

Hydrothermal liquefaction (HTL), which uses sub- and super-critical water as a reaction media, was used in this study to chemically recycle post-consumer recycled linear low-density polyethylene (PCR-LLDPE). In this study, it is hypothesized that PCR-LLDPE with lower molar mass can be hydrothermally depolymerized into oil with higher oil yield at the same reaction condition than those with higher molar mass. To examine this hypothesis, PCR-LLDPE with different molar mass ranges (7.20E+03-1.29E+05 g/mol) was converted into oil via HTL at 425°C with a reaction time of 2 h, a pre-determined reaction condition that can fully convert LLDPE into oil. The product yields validated our hypothesis—the lower the molar mass of PCR-LLDPE, the higher the oil yield. More specifically, when the molar mass is within the range of 7.20E+03-1.36E+04 (g/mol), a comparable oil yield (86-90 %) was obtained. However, when the molar mass increased from 1.06E+05 to 1.29E+05 (g/mol), a lower oil yield (73%) was realized. The chemical compositions of the oil converted from different molar mass ranges (7.20E+03-1.29E+05 g/mol) were also characterized, and found to mainly contain paraffin, olefins, cyclics, and aromatics. Additionally, it was found that PCR-LLDPE and the post-HTL solid residues contained metallic impurities such as Ca and Al that may participate in hydrothermal depolymerization of LLDPE, but more research is needed to elucidate the effect of these impurities on HTL of LLDPE.

505e Catalytic Deconstruction of Ethylene Vinyl Acetate Copolymer and Polyethylene Mixtures

Pedro Moura, Pavel Kots, Brandon Vance, Zoe Schyns, Sean Najmi, Zachary Hinton, Caitlin Quinn, Thomas H. Epps, III, LaShanda Korley, Dionisios Vlachos

We explore hydrogenolysis over ruthenium supported on zirconia (Ru/ZrO_2) and hydrocracking over platinum (Pt) supported on zeolites as an effective end-of-life strategy for ethylene vinyl acetate (EVA)—a widely used performance heat-sealant in hard-to-recycle multilayer packaging. For Ru/ZrO_2 hydrogenolysis, EVA reacts slower than low-density polyethylene (LDPE), and the catalyst deactivates due to carbonaceous deposits originating from polyenes generated *in-situ* during EVA thermal degradation. High temperatures can overcome catalyst deactivation; however, CH_4 yields are excessive. Hydrogenolysis of mixed EVA and linear low-density polyethylene (LLDPE), mirroring typical frozen food packaging formulations, results in comparable catalyst activity and CH_4 yield as the pure EVA resin. These findings showcase the inability of common hydrogenolysis catalysts to handle EVA and its associated mixtures

505f Exploring Chemical Looping Strategies for Waste Valorization to Generate High-Purity Commodity Chemicals

Ishani Karki Kudva, Shekhar Shinde, Tanay Jawdekar, Sudeshna Gun, Rushikesh Joshi, Sonu Kumar, Ashin Sunny, Liang-Shih Fan

Environmental degradation persists and is worsened by the accumulation of waste plastics, and biomass like bagasse, corn cob, etc., highlighting the urgent need for innovative solutions. Traditional methods of managing these materials include landfill disposal, incineration, and mechanical recycling. These methods have contributed to the degradation of ecosystems, air and water pollution, and worsening climate change. Plastic waste pollutes landfills and oceans, releasing toxins and disrupting marine life, while unregulated biomass disposal leads to deforestation, soil degradation, and greenhouse gas emissions.

The co-gasification of plastics and biomass provides various advantages, including a high H/C ratio of plastic that counterbalances the higher oxygen concentration in biomass. As a result, there is an increase in hydrogen yield, a reduction in tar generation, and an improvement in syngas quality. The use of biomass, which is renewable and carbon-neutral, complements the energy content of plastics in cogasification, resulting in a higher calorific value of the feedstock mixture and greater energy yields per unit of input. Including biomass in cogasification processes supports sustainability goals by decreasing dependence on fossil fuels and mitigating greenhouse gas emissions. Syngas generated from biomass can be considered carbon-neutral since the carbon dioxide released during combustion or utilization is compensated by the carbon dioxide absorbed during biomass growth.

The chemical looping process is a two-reactor system that combines plastic and biomass as feedstock. The feedstock in the first reactor, a moving bed reducer, and iron-based oxygen carriers move concurrently. The steam is added along with feedstock for char gasification. The oxygen carriers donate their lattice oxygen to carbon from the feedstock during this process, forming carbon monoxide (CO) and hydrogen (H_2). The result is high-purity syngas, i.e., $CO+H_2$, which can be upgraded to valuable chemicals in downstream processing. The second reactor, a fluidized bed combustor, regenerates the oxygen carriers using air and recirculates them back to the reducer. This process eliminates the need for expensive units like tar reformers, air separation, and acid gas removers.

The gasification process is highly endothermic; therefore, several approaches have been explored to counteract this and achieve an autothermal system, i.e., no external heat is required. One of these methods involves burning biomass in the combustor to supply heat to the process, while another involves co-feeding natural gas, waste plastics, and biomass into the system. This serves two purposes: creating an autothermal system due to the high heating value of natural gas, and using natural gas as a hydrogen source to produce higher-quality syngas. In the proposed strategies, the syngas purity is higher than 90%, and H_2:CO is approximately 2, which is required for downstream processing of valuable chemicals such as methanol. Compared to the existing indirect gasification of mixed plastic waste, the proposed strategies outperformed them. The indirect gasification process relies on oxygen from an air separation unit to burn natural gas, generating heat for plastic gasification while producing carbon emissions, either treated by a highly energy-intensive acid gas removal process or left untreated. This process was simulated using Aspen Plus software. The feasibility of the process was also investigated by conducting experiments on a 2.5 kW thermal bench scale unit. The coherence between the experimental and simulated data was verified, indicating that the process is feasible.

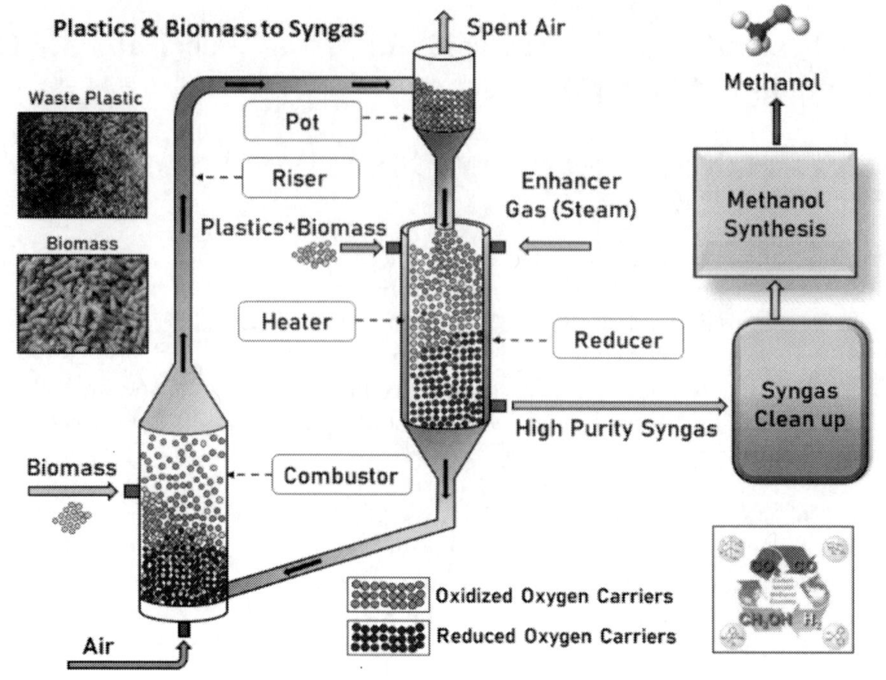
Plastics & Biomass to Syngas
Spent Air
Methanol
Waste Plastic
Pot
Riser
Enhancer
Gas (Steam)
Methanol
Synthesis
Biomass
Plastics+Biomass
Heater
Reducer
Syngas
Clean up
High Purity Syngas
Biomass
Combustor
Oxidized Oxygen Carriers
Reduced Oxygen Carriers
Air

505g Characterization of Plastic-Based Pyrolysis Oils: A Comparison between One-Dimensional Gas Chromatography and Two-Dimensional Gas Chromatography

Barbara Alejandra Perez, J.V. Jayarama Krishna, Hilal Ezgi Toraman

The U.S. produced 35.67 million tons of plastic waste in 2018, of which 3.09 million tons were recycled—while 26.97 million tons of plastic waste were landfilled, corresponding to 75% of all plastic waste produced in the U.S. in 2018. Polypropylene (PP), high-density polyethylene (HDPE), and polystyrene (PS) play a vital role across various applications, including durable goods, automotive components, containers, and packaging. The continuous use of PP, HDPE, and PS has resulted in a consequential surge in plastic waste, posing challenges in effective waste management and recycling. Among chemical recycling, pyrolysis stands out as a promising technique. It involves the decomposition of polymers by heating to a moderate temperature (400 – 600°C) under a non-reactive atmosphere in the absence or presence of catalysts. The pyrolysis oil can be further utilized by feeding it into steam cracker. However, pyrolysis oil resulted in intricate reaction networks influenced by sample characteristics, pyrolysis temperature, sample size, and carrier gas flow rate. These factors contribute to the complexity of reaction pathways, which may result in limitations related to kinetic measurements and transport phenomena if these factors are overlooked.

To evaluate the abovementioned parameters, it is necessary to have a comprehensive understanding of the pyrolysis product speciation data. Moreover, the speciation data for the pyrolysis of polymer are fundamental to assessing (micro-) kinetic models, which need to be validated with data obtained under well-defined pyrolysis conditions and a kinetically controlled regime. Traditionally, pyrolysis studies have employed one-dimensional gas chromatography (GC) coupled to mass spectrometry (GC–MS) or flame ionization detector (GC–FID). Nevertheless, peak resolution is compromised when GC is used for complex matrices such as pyrolysis oils. While GC may result in complicated chromatograms with overlapping peaks failing the identification and quantification of pyrolysis products, two-dimensional gas chromatography (GC×GC) offers enhanced resolution, achieved by a two-column combination with different separation mechanisms connected in series through a modulator. GCGC can be coupled simultaneously to FID and MS, such as a time-of-flight mass spectrometer (GC×GC–FID/TOF–MS).

This work examined the product distribution of plastic-based pyrolysis oil by using two analytical techniques: GC–MS and GC×GC–FID/TOF–MS. The pyrolysis of the single polymers (i.e., PP, HDPE, and PS) and their co-pyrolysis were conducted in a micropyrolyzer coupled to GC–MS and GC×GC–FID/TOF–MS at a pyrolysis temperature of 500 °C. The total yield for the pyrolysis of for PP and PS was 43.9 ± 0.7 and 86.9 ± 2.3 wt.%, respectively. Meanwhile, the co-pyrolysis of PP-PS yielded 64.0 ± 1.4 wt.%. GC×GC identified up to 6 times more peaks compared to GC, improving the identification of pyrolysis products with higher polarity (i.e., aromatics) using a normal column combination. Moreover, new pyrolysis products were identified using GC×GC. The new products identified corresponded diolefins, cyclic hydrocarbons, and aromatic hydrocarbons. For instance, new products identified using GC×GC in the pyrolysis of PP yielded up to 1.3 wt.%. These chemical classes may be produced by secondary reactions induced by transport effects as well as higher vapor residence times; however, GC overlooks these chemical classes due to inferior resolution compared to GC×GC. It is worth noting that under the low gas residence times investigated in this study, secondary reactions were not prominent. Nonetheless, the high resolution and sensitivity of GC×GC facilitated the detection of these components. Besides, a clear enhancement in the identification of the pyrolysis products occurred when GC×GC was used to analyze the pyrolysis oil products from the co-pyrolysis. For example, GC×GC was able to identify 38 new products from the co-pyrolysis of PP-PS, which corresponded to a yield of 0.5 ± 0.07 wt.%. On the other hand, these products were not identified using GC due to low intensity and co-elution of the GC chromatogram. Considering the possible increased extent of secondary reactions at larger scale reactors, utilization of GC×GC analysis could become more essential. The findings of this work clearly show that a comprehensive knowledge of the product distribution of plastic pyrolysis oils is fundamental to acknowledging unwanted by-products and, therefore, assessing precise (micro-) kinetic models.

505h Polymer Scission in Catalytic Nanopores: Role of Polymer Dynamics in Upcycling

R Bharath Venkatesh, Jon Bingaman, Susannah Scott, Rachel Segalman

Addressing the issue of plastic waste pollution requires innovations in end-of-life handling of polymer-based products. Upcycling processes that take advantage of catalytic processes to convert waste polymers into commodity products are often limited by low conversion rates and poor selectivity. One way to improve the scalability of such processes is to employ nanoporous solid catalysts that cleave the polymer backbone selectively at high rates. However, the entropic penalty for polymers to enter catalytic nanopores imposes severe transport limitations on large polymers to access these active sites. Yet, several studies report that the physical attributes of nanopores such as pore size and lateral dimensions can control the selectivity of the final products. Batch-scale investigations of catalytic polymer scission are limited in their elucidation of physical principles governing polymer deconstruction, especially pore-level macromolecular reaction mechanisms. To develop an understanding of polymer reactivity decoupled from transport processes, we utilize dielectric spectroscopy to monitor the scission of polymers confined inside Anodized Aluminum Oxide (AAO) nanopores. AAO nanopores are used as model catalytic pores with uniform geometry and acid sites, capable of chain scission, can be localized along the pore walls. Change in the segmental dynamics of reacting polymer chains, obtained from dielectric spectra, serve as an indicator of the progress of the reaction and help map out the effective reaction rates inside nanopores. We demonstrate the validity of this technique by tracking the degradation of poly (propylene carbonate) inside AAO nanopores. A model polyolefin, poly (ethylene-alt-propylene), is then used along with nanopores of varying morphology and active site density to elucidate the factors influencing polymer reactivity inside catalytic nanopores.

361a Convergent Plastics Recycling Research and Innovation Based at the NYS Center for Plastics Recycling Research and Innovation

Amit Goyal

This talk will provide an overview of plastics recycling research being conducted at the NYS Center on Plastics Recycling Research & Innovation at the University at Buffalo. The Center focuses on addressing a range of contemporary issues related to plastics recycling, with the goal of advancing plastics recycling in NY state and beyond. The Center is based on ***multidisciplinary, convergent research*** and involves faculty and students from several schools and colleges spanning the disciplines of *operations research*, *environmental engineering*, *chemistry*, *chemical engineering*, *materials science*, *economics, physics, data science* and *communications*. Ongoing Tasks include: **1:** Mapping the reverse supply for plastics in NYS and deep dive into the structure of the collection, disposal, and recycling industries; **2:** Improving the efficiency of businesses involved in collecting and processing plastics; **3:** Plastics in NYS agriculture/food and medical industry; **4:** Further understand contemporary attitudes toward recycling and current recycling behaviors and to design and test campaigns to increase recycling knowledge and minimize contamination in recycling; **5:**Microplastics in Natural Environments; **6:** Advanced High-Speed Plastic Sorting Technique by Molecular Contrast Infrared Imaging for advancing plastics recycling; **7:** Determining percentage of recycled plastic content in plastic products via transformative new technologies. The talk will provide a summary of results obtained from these projects.

Acknowledgements: The NYS Center on Plastics Recycling Research & Innovation at the University at Buffalo acknowledges funding and support from NYDEC.

361b Coupling of Ammonolysis and Pyrolysis to Upcycle Challenging Multilayer Plastic Films

Daniel Kulas, Sulihat Aloba, Lindsay Putman, Sarvada Chipkar, Adrian Noecker, JoAnn Henry, Emily Leverance, Colin Meinholz, Kevin Garland, Steve Techtmann, Rebecca Ong, David Shonnard

Multilayer plastic films are complex materials containing distinct layers of materials including many types of plastics, aluminum, and paper. Each layer is added to confer a property to the film, such as a gas/vapor barrier, mechanical strength, or heat tolerance. Recycling technologies for multilayer materials are all at low readiness levels, with several technical hurdles still to be overcome for commercialization. In this work, we introduce a novel sequential chemical upcycling technology for multilayer materials, sequential ammonolysis and fast pyrolysis, which were used to process multilayer military meals ready-to-eat (MRE) food packaging with 4 distinct layers: 25-35% polyethylene (PE), 50-60% polyethylene terephthalate (PET) and polyamide (PA), and 10-20% aluminum. Being able to recycle MRE packaging would eliminate the costs and hazards associated with disposing of waste at a military base and potentially allow the military to recover value from a waste stream. Ammonolysis completely solubilized the PET and PA, converting them into monomers that were used as a feedstock for microbial growth to make single-cell protein powder. The PET and PA breakdown products were measured using HPLC to confirm complete solubilization of the PET and PA fractions in the mixed waste. The residual solids were subjected to pyrolysis to completely convert PE into wax, liquid, and gas products, which could be used in gaseous and liquid fuel blends, for power generation, and as lubricants, with the aluminum being recovered as a component of pyrolysis char. The pyrolysis products from the residual solids have comparable product quality metrics, including GC-MS, FTIR, CHNO, and metals analysis, to that of pure PE pyrolysis. Additionally, results will be reported on the ability of microbes to grow on the PET and PA breakdown products to produce single-cell protein. Preliminary results from culturing done with PET-degrading consortia and isolates reveal good growth on deconstructed products derived from brown and green MRE bags with cultures reaching optical densities in the 0.5 – 1 range as measured using spectrophotometry (600 nm).

361c Bio-Based and Biodegradable Plastics - When and Where Do They Make Sense?

Rachael Rothman, Stuart Walker, Fanran Meng, Tony Ryan

Bio-based and biodegradable plastics offer interesting alternatives to traditional fossil based plastics, however it is not as simple as just substituting one for the other. Producing bio-based plastics uses a lot of energy, with associated environmental impacts, and biodegradable plastics must degrade in the environments they find themselves and not contaminate recycling streams.

Here we discuss some of the challenges associated with life cycle assessment of bio-based and biodegradable plastics, then we use life cycle assessment to compare pairs of similar polymers, for example bio-based and fossil based PET, and Polystyrene and PLA. The assessment takes into account the predicted decarbonisation of the electricity grid and the use of alternative heating sources to replace natural gas. The comparisons allow a 'cross-over' to be evaluated - the year in which the bio-based polymer has a lower global warming potential than the fossil-based polymer.

The bio sources used to make bioplastics take in carbon dioxide during growth. When the bioplastic reaches end of life this carbon is most often released to the atmosphere, either through composting, incineration or landfill. If instead bioplastics were made to be durable, they would effectively lock in carbon for a long period of time and become a route for carbon sequestration. Over 1 gigatonne of carbon could be sequestered in plastics every year1. The sequestration potential of plastics will be discussed, highlighting the opportunities and challenges.

[1] Ryan, A.J., Rothman, R.H. Engineering chemistry to meet COP26 targets. Nat Rev Chem 6, 1–3 (2022). https://doi.org/10.1038/s41570-021-00346-6

361d Health Effects of Microplastics and Nanoplastics on Humans

Jeffrey Seay, Teresa Murray, Joan G. Lynam, Robert Peters

Plastics are ubiquitous in the environment. There have been increasing reports of adverse health effects reported on humans, mammals, and the ecosystem. This paper addresses the impact on human health associated with microplastics and nanoplastics, based on reported health effects provided in the technical and scientific literature.

361e Hydrothermal Liquefaction of Cyanobacteria and Ocean-Bound Polyethylene Plastic Waste

Daniel Lachos Perez, Andrew Halloran, Kalsoom Jana, Grace Chen

Plastics are pervasive environmental contaminants polluting the planet. Microplastics are plastic particles < 5-mm in size. Due to their small size, microplastics may be ecotoxic and accumulate cyanobacterial toxins released by harmful algal bloom (HAB). Given the pervasiveness of microplastic pollution and HAB, exploring technologies that can simultaneously handle microplastics and HAB while converting them into valuable products is crucial. Therefore, in this study, model HAB (Spirulina – SP) and ocean-bound polyethylene plastic (PE) waste were coprocessing using hydrothermal liquefaction (HTL). The initial screening tests for optimizing HTL reaction conditions included reaction times (0.5, 1, and 2h), SP solid concentration (10 and 25 wt.%), and temperature (300, 350, and 400 °C). After HTL reactions, four streams of products including the biocrude oil, aqueous phase, gas, and solid residue were produced and their respective yields were quantified. The chemical composition of the biocrude oil was also analyzed by using gas chromatography mass spectrometry (GC-MS). The screening results revealed that the highest oil yield (26.67 wt.%) was achieved at 2h, 10 wt.% SP, and 400 °C, with GC-MS analysis revealing the major chemical composition of biocrude oil including alkanes, alkenes, phenols, amides, aromatics and heteroatoms of nitrogen compounds. To examine potential synergies in co-converting PE and SP, additional HTL tests will be conducted using different ratios of SP and PE (SP:PE=1:0, 3:1, 1:1, 1:3, 0:1) under the down-selected optimal condition (2h, 10 wt.% SP, and 400 °C). The compositions of the biocrude oil obtained will be analyzed by GC-MS and be used to guide follow-up studies using field samples (microplastics and HAB). Fundamental insights learned from this study will be useful to guide the design of energy-efficient microplastic and HAB co-mitigation methods in future studies.

385a Preliminary Techno-Economic Analysis of PET Electro-Reforming Process

Hyeonseo IM, Woochang Jeong, Jonghwan Oh, Yong Sang Lee, Heehyun Choi, Janggeun Lee, Jiyong Kim

This study aims to examine the techno-economic-environmental criteria of the Polyethylene terephthalate (PET) electro-upcycling process. To achieve this goal, we developed PET electro-reforming process involving enzymatic depolymerization and electrochemical technologies to produce value-added chemicals (e.g., terephthalic acid, glycolic acid, hydrogen) from recycled PET flakes as a feedstock. With the developed process, we evaluated the economics of the process based on the minimum selling price (MSP) of terephthalic acid and glycolic acid and environmental impact of the process based on net CO_2 equivalent emission (NCE). Also, we performed sensitivity and uncertainty analysis to identify the feasibility of the process with the fluctuation on major parameters. As a result, proposed process demonstrates economic feasibility with low NCE. Consequently, PET electro-upcycling process can be regarded as an effective pathway for chemical upcycling of PET.

PET is indispensable product which is widely used for food packaging, disposable medical equipment, and daily necessities [1]. With its high usability and great physical properties, annual worldwide market volume of PET has increased to 25.47 million metric tons (MMT) and is expected to increase continuously [2]. However, this substantial production volume generates significant waste, leading to persistent pollution to the ecosystem. In this regard, developing effective, economic technology for recycling is crucial to addressing the plastic waste problem. Recently, PET electro-reforming technology has emerged as a solution with considerable advantages (e.g., controllability of reaction rates by managing applied voltage, wide adaptability due to modular electrolyzer designs, and compatibility with depolymerization process of PET) [3]. While active research is being conducted at the lab scale, the economic and environmental impacts of this technology at an industrial scale remain uncertain. Thus, the development of PET electro-reforming process on an industrial scale is necessary to evaluate the feasibility of the process.

The goal of this study is to examine the techno-economic-environmental performance of PET electro-reforming process. To achieve this goal, we developed the PET electro-reforming process which selects recycled PET flakes as a feedstock to produce Terephthalic acid (TPA), Glycolic acid (GA) as a main product by employing several technologies. The technologies include reaction and conversion technologies (e.g., enzymatic depolymerization, electrolysis), component separation (e.g., acidification, filtration, membrane), and product purification (e.g., centrifugation, flash drum, distillation column). The process flow diagram of this study is shown in the Figure. In this study, we evaluated economic feasibility of the PET electro-reforming process using minimum selling price (MSP) of TPA and GA. The MSP includes total capital investment (TCI), total operating costs (TOC), and profits from by-products (i.e., H_2 and K_2SO_4). Expenses related to electrolysis (e.g., electrolyzer capital cost, BoP capital expenditures and stack replacement interval) are estimated based on the H2A production case study report [4]. The mass and energy information along with sizing and cost information is obtained by process simulation using Aspen plus V.12. To evaluate environmental impact of the process, net CO_2 equivalent emissions (NCE) is used. In this process, we only consider indirect CO_2 emissions related to utility consumption, as there are no direct CO_2 emissions. We also performed the sensitivity analysis to investigate major cost and environmental drivers of the process. Finally, we performed uncertainty analysis based on Monte Carlo (MC) simulation to determine the feasibility of the process with the cost fluctuation on major cost drivers [5]. From the MC simulation, a probability range of MSP is observed, so that economic competitiveness of the process is identified.

In this study, we developed a PET electro-reforming process and evaluated the process from a techno-economic-environmental perspective. Compared to the current market prices of the main products, the MSPs of TPA and GA from our developed process show negligible differences. The recycling of electrolyte and unreacted reactant enhances the overall efficiency of the process. Major cost contributors that significantly influence the MSP include the price of PET, GA, K_2SO_4, and KOH as well as the interest rate and the faradaic efficiency of electrolysis. Through sensitivity and uncertainty analysis, we have also revealed that process competes economically within the range of cost fluctuations. Additionally, NCE of the process is low due to low energy consumption for heating and the use of renewable electricity. In addition, the early-stage evaluation of the PET electro-reforming process provides valuable insights into the economic and environmental feasibility of the technology, as well as its potential for future utilization.

Reference

[1] Rahimi, A.R., and J.M. Garciá, "Chemical recycling of waste plastics for new materials production," Nature Reviews Chemistry 1, 2017..

[2] Bescond, A.-S., and Pujari, A. (2020). PET polymer. In Chemical Economics Handbook (IHS Markit), p. 32. https://ihsmarkit.com/ products/chemical-economics-handbooks. html.

[3] Jouny, M., W. Luc, and F. Jiao, "General Techno-Economic Analysis of CO2 Electrolysis Systems," Industrial and Engineering Chemistry Research, 57 (6), pp. 2165–2177 (2018).

[4] D. Peterson, J. Vickers, and D. DeSantis, "Hydrogen Production Cost from PEM Electrolysis - 2019", (2020).

[5] Do, T.N., H. Kwon, M. Park, C. Kim, Y.T. Kim, and J. Kim, "Carbon-neutral hydrogen production from natural gas via electrified steam reforming: Techno-economic-environmental perspective," Energy Conversion and Management, 279, (2023).

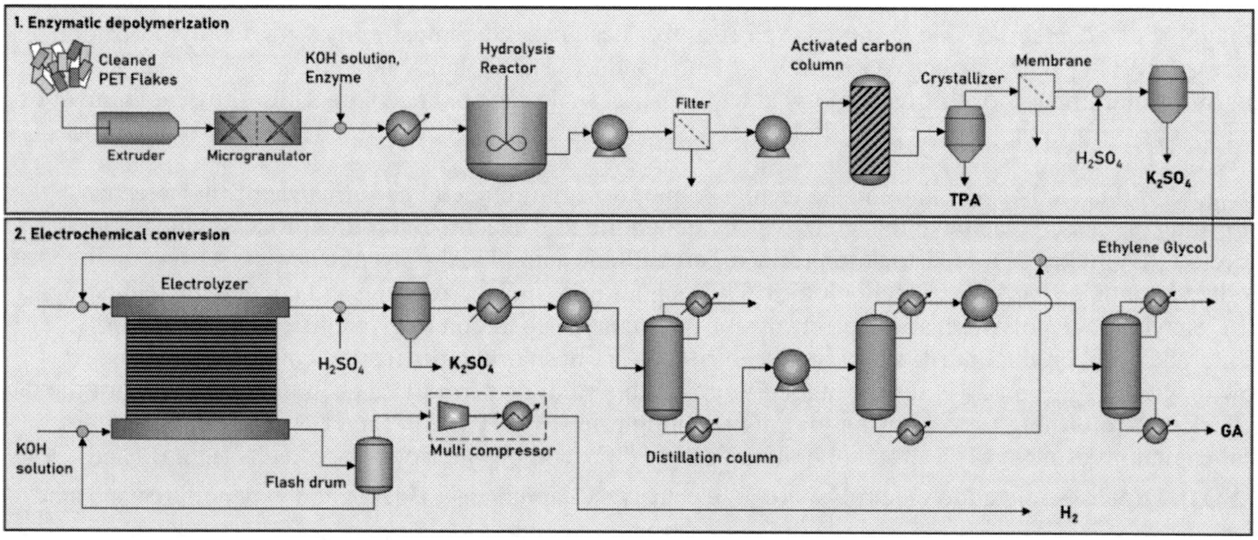

385b Characterizing Polyvinyl Chloride Interactions with Common Commercial Additives in Traditional and Bioderived Solvents

Feranmi Olowookere, Gabriel D. Barbosa, Heath Turner

In this study, we conduct a comprehensive investigation into the interactions between polyvinyl chloride (PVC) and its common commercial additives (plasticizers, heat stabilizers, flame retardants, etc.), with a focus on exploring alternative, bioderived solvents for processing PVC waste. Using molecular simulations, we evaluate the performance of five bioderived solvents and compare them against other traditional solvents, such as tetrahydrofuran (THF) and ethyl acetate (ETA). We examine the impact of the different solvents on the physical and mechanical properties of PVC, as well as their propensity to solvate PVC and its additives. Our findings suggest that bioderived solvents may provide comparable, if not superior, PVC solvation performance relative to traditional solvent formulations, highlighting these compounds as potential eco-friendly alternatives. Future experimental studies are planned to validate these findings and assess the practicality of these bioderived solvents, considering factors such as cost, synthesis feasibility, and toxicity.

385c Catalytic Pyrolysis of Disposable Masks: Adjusting Nickel Content Ratios in Metal Oxide Catalysts for Hydrogen-Rich Gas and Carbon Nanomaterial Quality

Do Hyun Lee, Ana Gabriela Mendoza Burgos, Hongryeol Kim, Jaewook Myung

The widespread use and disposal of disposable masks have raised social and environmental concerns. To address this issue, we explored the application of non-noble metal catalysts, specifically metal oxides ($LaNiO_3$, Ni/Al_2O_3, Ni/Zeolite 4A), in the catalytic pyrolysis of disposable masks. Our study focused on impregnated catalysts (Ni/Al_2O_3, Ni/Zeolite 4A), along with $LaNiO_3$. We found that these catalysts yielded comparable outcomes in terms of hydrogen concentration in the product gas and the structural purity of the resulting carbon nanomaterial. This suggests that the nickel content ratio in the catalyst plays a crucial role in the production of hydrogen and carbon nanomaterials during the catalytic pyrolysis of disposable masks. These findings indicate the potential for catalytic pyrolysis to support the sustainable production of carbon nanomaterials and clean gaseous fuels, offering a practical solution for transforming waste masks.

385d De-Plasticization of Expanded Polystyrene (EPS) Wastes into Value-Added Carbon Materials

Dipendu Saha, Hippolyte Grappe

Expanded polystyrene (EPS) or foam #6, popularly known as "styrofoam", is widely used in various household and industrial applications, like home and appliance insulation, packaging materials, food containers and insulators, road bank stabilization system, and many others. EPS is one of the highest single-use plastics with minimum recycling value. Approximately 1% of EPS is recycled, 99% is accumulated in the solid waste. EPS-based materials are responsible for 30% of total landfill waste, and in the U.S. alone, about 1,369 tons of EPS is accumulated in the waste every year. EPS is not bio-degradable, and it causes hazardous conditions for both land and aquatic life. In the environment, it may slightly depolymerize to produce traces of styrene which is a suspected carcinogen. As 90% of EPS is air, it is very light, voluminous, and inexpensive. These reasons make the traditional recycling of EPS unprofitable and unacceptable by most recycling plants. Therefore, upcycling EPS-based objects into value-added materials may provide a large impetus toward waste minimization and the economy of EPS.

We have developed the technology to sustainably de-plasticize EPS into (i) Carbon molecular sieve (CMS), (ii) Activated carbon, and (iii) Synthetic graphite. Unlike other plastics, when pristine EPS is pyrolyzed or carbonized, there remains to no carbon or char yield. Our technology will first convert into a stabilizing form, which will yield about 90% carbon upon pyrolysis with respect to pristine EPS. There is an extremely large pool of applications of carbon molecular sieves and activated carbons, including water purification, air purification, gas separation, mercury control, electrodes, supercapacitors, fuel cells, pharmaceutical product purification, medical applications for stomach detoxification, gas masks, and many others. Synthetic graphite is used as an anode of Li-ion batteries, furnace electrodes, coatings, conductive films, and others. Converting EPS wastes to these value-added carbon materials will not only help to remove the pollution caused by EPS but also help the EPS from being banned in the future. In addition, there is a supply chain problem of all these carbon materials and a significant portion of the domestic demands of the United States is met by imported materials from foreign countries. Therefore, domestic manufacturing of these carbon materials will partly solve the current supply chain problem.

385e Techno-Economic Analysis and Life Cycle Assessment of Municipal Plastic Waste Incineration, Gasification, and Pyrolysis Oil Production By Chemical Process Simulation

Soyoung Kang, Sanghun Lee

With the emergence of the transition to a circular economy to achieve net-zero emissions, it is important to determine how to dispose of Municipal plastic Waste (MPW). There are two methods for recycling MPW: gasification and production of pyrolysis oil. In the gasification process, syngas is produced through three steps: pretreatment, decomposition, and gasification. Also, MPW can be converted to liquid hydrocarbon fuel through pyrolysis consisting of four steps: pretreatment, pyrolysis, hydrogen production from pyrolysis gas, and hydrodeoxygenation using pyrolysis oil. These processes convert waste into valuable resources, aligning with the goals of a circular economy through chemical recycling. Most of existing research has focused on specific waste management technologies. Therefore, this study conducted simulations of incineration, gasification, and pyrolysis oil production using Aspen Plus software to process 1000 tonnes of MPW daily. For techno-economic analysis, profits were calculated by considering disposal costs and revenue from final products based on the Aspen Plus model. Additionally, a Life Cycle Assessment (LCA) was performed to quantify greenhouse gas emissions using the GREET model. The gasification process has the higher efficiency of 82.1% compared to the pyrolysis oil production process at 60.7%, resulting in profits of $782.36 and $338.13 per tonne of disposed MPW, respectively. Although the cost for the gasification process is higher, it generates greater profits due to the large-scale hydrogen production. LCA indicates that pyrolysis oil production has the lowest Global Warming Potential (GWP) among the processes, with values of 2300.70 kg CO_2 – eq/tonne of MPW for pyrolysis oil production, 3397.25 CO_2 -eq/tonne of MPW for gasification, and 4847.10 kg CO_2 -eq/tonne of MPW for incineration. This suggests that the environmental impact can be reduced by up to 50% when plastic waste is recycled instead of incineration. Overall, the gasification process is a complex and energy-intensive process. However, it has potential for hydrogen production. The pyrolysis oil production process transforms MPW into hydrocarbon fuel at low temperatures and has a low environmental footprint. Further exploration of gasification with Carbon Capture and Storage (CCS) and oil production with advanced refining systems can be suggested for future research on waste plastic management.

385f Techno-Economic Analysis of Inherently Recyclable All-Polyester Multilayer Plastics

Dilara Goreke, Francesca Pierobon, Kevin Simmons

Owing to their affordability and unrivaled characteristics such as flexibility, durability, lightness, and longevity, the production of multilayer plastics (MLPs) experienced a rapid increase, reaching more than 100 million tons of flexible and rigid MLPs production per annum and representing over 30% of all plastics produced [1,2]. These MLPs comprise several layers of different materials, each selected for their specific physical, chemical, and mechanical properties [3]. This complexity, however, poses significant challenges for current recycling systems, resulting in most of these materials ending up in landfills.

In this context, we present a comprehensive techno-economic analysis (TEA) for the production and mechanical recycling of all-polyester MLPs that are inherently recyclable because their layers are chemically and mechanically compatible. We conduct a discounted cash flow analysis to determine the minimum selling price of different film structures composed of various polyester resins that can meet or surpass the functionality of commercial nonrecyclable 5–12-layer MLPs. We compare these prices with conventional benchmarks and conduct a sensitivity analysis of key economic parameters to understand their impact on the process economics. Our study identifies economically viable film structures and recycling technologies that contribute to sustainable waste management practices.

References:

[1] Collias, Dimitris I., Martin I. James, and John M. Layman (Nov. 2021). "Introduction—Circular Economy of Polymers and Recycling Technologies". In: ACS Symposium Series. American Chemical Society, pp. 1–21. doi: 10.1021/bk-2021-1391.ch001.

[2] Lithner D, Larsson Å, Dave G. Environmental and health hazard ranking and assessment of plastic polymers based on chemical composition. Science of The Total Environment. 2011;409(18): 3309-24.

[3] Walker TW, Frelka N, Shen Z, Chew AK, Banick J, Grey S, Kim MS, Dumesic JA, Van Lehn RC, Huber GW. Recycling of multilayer plastic packaging materials by solvent-targeted recovery and precipitation. Science Advances. 2020;6(47):7599.

385g Sulfuric Acid for Chemical Recycling of Polycarbonate in Water

Hossein Abedsoltan

This study is centered on the chemical recycling of poly (bisphenol-A carbonate) (PC) pellets through hydrolysis, where concentrated sulfuric acid is utilized as a catalyst in a water medium. The impact of varying sulfuric acid concentrations and reaction temperatures on the hydrolysis rate of PC was investigated. Experiments were conducted to observe the conversion rate of PC over time with sulfuric acid concentrations ranging from 10M to 16M and temperatures between 110°C and 150°C. The kinetics of the PC hydrolysis process were modeled using a pseudo-first-order reaction framework, through which specific reaction rate constants were determined. The overall activation energy for the hydrolysis of PC was calculated using the Arrhenius equation. The recovery of the catalyst (sulfuric acid) post-hydrolysis through distillation, its characterization via titration, and its reapplication in subsequent hydrolysis experiments to assess catalyst reusability were also explored. It was found that sulfuric acid can be maintained in its acidic state and is effective as a catalyst for up to five cycles of reuse. The reduction in the size of the PC due to hydrolysis was examined as well. Moreover, a mechanism for the hydrolysis of polycarbonate in the presence of an aqueous solution of sulfuric acid was proposed.

385h Molecular Weight Fractionation of Polyvinyl Chloride (PVC) in Green Solvent Systems

Jaewoo Choi

Polyvinyl chloride (PVC) is a widely used polymer with diverse applications, ranging from construction materials to medical devices to consumer products. Fractionation (i.e., the separation of polymer chains based on molecular weights) can be applied to PVC as a means of removing lower molecular weight content and modifying overall properties. In this study, we investigated the fractionation of PVC in using mixtures of two green solvents: methanol and acetone. We systematically examined the effects of solvent composition on the fractionation efficiency and yield of commercial PVC from a major US supplier. Gel permeation chromatography (GPC) was used to characterize the molecular weight and polydispersity of each fraction and elucidated the influence of solvent composition on the fractionation process. Our findings reveal significant opportunities to fractionate PVC with simple and environmentally friendly solvent compositions. This study contributes to a deeper understanding of PVC fractionation mechanisms and provides valuable insights for optimizing PVC processing and applications.

385i Converting Waste Plastics into Hydrocarbon Products: Exploring the Economic Potential of Hydrocracking Process

Elizabeth Aigaje, Rui Shi

Plastics have been integral to human life for more than a century, with their widespread use spanning the globe. They are versatile, flexible, lightweight, and low-cost, which make plastics suitable materials for numerous applications, including packaging, construction, textiles, and more [1], [2]. In 2022, global plastic production was 400.3 million metric tons, marking a 1.6% increase from 2021 [3]. On average, polyolefins constitute 57% of the plastics produced, followed by PVC at 12% , with PET, PUR and PS each accounting for approximately 10% each one [4], mirroring a similar trend in the composition of plastic waste. The increasing rate of plastic production, coupled with the short life span of most plastics, their non-biodegradability nature, and inadequate end-of-life management, have result in the accumulation of plastics in Earth's soil, water, and air, becoming a persistent environmental challenge.

Currently, the primary method of plastic waste disposal in the USA is landfilling, followed by 9% of plastic incinerated and only 5% recycled [5]. Addressing the challenge of plastic waste management necessitates the development of large-scale recycling technologies. Mechanical recycling involves reprocessing plastic to manufacture new products. However, the properties of plastic degrade in each recycling cycle On the other hand, chemical recycling emerges as a promising alternative that adds value to plastic waste by converting them into a spectrum of chemical compounds, monomers and fuels [6].

Different chemical recycling processes have been studied such as pyrolysis, hydrothermal liquefaction (HTL), solvolysis, hydrogenolysis, hydrocracking and others [7]. Among these, hydrocracking stands out as a well-established process in the refinery industry that has recently gained attention for its efficacy in decomposing plastics, particularly for polyolefins [8], [9], [10]. In hydrocracking, heavy molecules are transformed into lighter molecules through carbon-carbon cleavage, similar to pyrolysis, but under the presence of hydrogen. Noteworthy advantages of hydrocracking include low operation temperatures, high production of alkanes over olefins, and low coke formation. Moreover, hydrocracking yields high-quality liquid oil without the need of several separation steps [11].

The development and adoption of chemical recycling technologies underscore the critical necessity for economic analysis to guarantee their long-term viability at an industrial scale. However, compared to other chemical recycling technologies, only a limited number of studies have reported the techno-economic analysis (TEA) for the hydrocracking of plastic waste [12], [13]. TEA typically employs discounted cash flow analysis to determine metrics such as minimum selling price (MSP) of the desired product, net present value (NPV), internal rate of return (IRR) among others. These indicators serve to evaluate and compare the economic feasibility of the technology. Consequently, we propose to conduct a comprehensive techno economic analysis for the hydrocracking of plastic waste at its early stages, accounting for different scenarios. Specifically, we aim to assess the impact of the catalyst performance (e.g., yield, conversion) and feedstock pricing on the economic viability of hydrocracking. Our goal is to deliver a robust analysis delineating the conditions wherein the chemical recycling process maintains feasibility. This will pave the way for future research efforts to converge effectively, fostering the creation of viable technology.

References

[1] A. J. Martín, C. Mondelli, S. D. Jaydev, and J. Pérez-Ramírez, "Catalytic processing of plastic waste on the rise," *Chem*, vol. 7, no. 6, pp. 1487–1533, Jun. 2021, doi: 10.1016/j.chempr.2020.12.006.

[2] W. Tian *et al.*, "Microplastic materials in the environment: Problem and strategical solutions," *Progress in Materials Science*, vol. 132, p. 101035, Feb. 2023, doi: 10.1016/j.pmatsci.2022.101035.

[3] Statista, "Annual production of plastics worldwide fromo 1950 to 2022." Accessed: Feb. 08, 2024. [Online]. Available: https://www.statista.com/statistics/282732/global-production-of-plastics-since-1950/#:~:text=The%20worldwide%20production%20of%20plastics%20reached%20a%20staggering,the%20continued%20growth%20in%20production%20year%20after%20year

[4] R. Geyer, J. R. Jambeck, and K. L. Law, "Production, use, and fate of all plastics ever made," *Sci. Adv.*, vol. 3, no. 7, p. e1700782, Jul. 2017, doi: 10.1126/sciadv.1700782.

[5] A. Milbrandt, K. Coney, A. Badgett, and G. T. Beckham, "Quantification and evaluation of plastic waste in the United States," *Resources, Conservation and Recycling*, vol. 183, p. 106363, Aug. 2022, doi: 10.1016/j.resconrec.2022.106363.

[6] F. Perugini, M. L. Mastellone, and U. Arena, "A life cycle assessment of mechanical and feedstock recycling options for management of plastic packaging wastes," *Environ. Prog.*, vol. 24, no. 2, pp. 137–154, Jul. 2005, doi: 10.1002/ep.10078.

[7] O. D. Bozkurt, S. T. Okonsky, K. Alexopoulos, and H. E. Toraman, "Catalytic conversion of SPW and products upgrading," in *Advances in Chemical Engineering*, vol. 60, Elsevier, 2022, pp. 117–168. doi: 10.1016/bs.ache.2022.09.006.

[8] L. Li *et al.*, "Converting Plastic Wastes to Naphtha for Closing the Plastic Loop," *J. Am. Chem. Soc.*, vol. 145, no. 3, pp. 1847–1854, Jan. 2023, doi: 10.1021/jacs.2c11407.

[9] S. Liu, P. A. Kots, B. C. Vance, A. Danielson, and D. G. Vlachos, "Plastic waste to fuels by hydrocracking at mild conditions," *Sci. Adv.*, vol. 7, no. 17, p. eabf8283, Apr. 2021, doi: 10.1126/sciadv.abf8283.

[10] Z. Qiu *et al.*, "A reusable, impurity-tolerant and noble metal–free catalyst for hydrocracking of waste polyolefins," *Sci. Adv.*, vol. 9, no. 25, p. eadg5332, Jun. 2023, doi: 10.1126/sciadv.adg5332.

[11] D. Munir, M. F. Irfan, and M. R. Usman, "Hydrocracking of virgin and waste plastics: A detailed review," *Renewable and Sustainable Energy Reviews*, vol. 90, pp. 490–515, Jul. 2018, doi: 10.1016/j.rser.2018.03.034.

[12] B. Hernández, P. Kots, E. Selvam, D. G. Vlachos, and M. G. Ierapetritou, "Techno-Economic and Life Cycle Analyses of Thermochemical Upcycling Technologies of Low-Density Polyethylene Waste," *ACS Sustainable Chem. Eng.*, vol. 11, no. 18, pp. 7170–7181, May 2023, doi: 10.1021/acssuschemeng.3c00636.

[13] O. Olafasakin, J. Ma, V. Zavala, R. C. Brown, G. W. Huber, and M. Mba-Wright, "Comparative Techno-economic Analysis and Life Cycle Assessment of Producing High-Value Chemicals and Fuels from Waste Plastic via Conventional Pyrolysis and Thermal Oxo-degradation," *Energy Fuels*, vol. 37, no. 20, pp. 15832–15842, Oct. 2023, doi: 10.1021/acs.energyfuels.3c02321.

385j Synthesizing Highly Crystalline Graphite Powder for Lithium-Ion Battery Anodes from Bulk Polyethylene Waste

Yuan Gao, Ngoc Tien Huynh, Viet Hung Pham, Congjun Wang, Christopher S. Matranga, Ki-Joong Kim

Upcycling plastic waste into graphite solves an issue controlling supply chain stability and creating a new path for graphite manufacturing. However, synthesizing graphite from plastic waste, particularly polyethylene (PE) plastics, remains a significant challenge because it requires a stabilization process that has not been studied as a pathway for large-scale upcycling of bulk plastic wastes. In this work, we demonstrated a straightforward method of air processing yielding a carbonaceous material (or carbon char) from PE bulk plastic waste using solid additives. The air processing developed in this study overcomes the oxygen diffusion bottlenecks for processing bulk PE waste, therefore achieving char yield over three orders of magnitude with values as high as 20 wt.% from bulk PE plastics, compared to without using solid additives. We also proved that thermally stabilized PE materials can be converted into highly crystalline flake graphite via low-temperature catalytic graphitization. The assembled anode PE-derived graphite showed comparable electrochemical performance, such as reversible rate performance and long-term cyclic stability, to the current-use battery-grade graphite, thus providing a scalable method to upcycle PE plastic waste into high-value carbon materials and is adaptable to practical applications.

385k Upcycling Plastic Wastes to Fuels and Monomers Via Electrified Spatiotemporal Heating

Qi Dong

Nearly 9 billion metric tons of plastic have been produced in the human history, among which approximately 2/3 have become waste that can result in long-term negative environmental and ecological impacts. With the continuing heavy demand for plastics, it is urgent to develop efficient plastic recycling and upcycling strategies to ensure a sustainable future. While thermochemical plastic upcycling can generate more useful chemicals such as fuels and monomers compared to other methods, conventional thermochemical approaches generally suffer from low product selectivity and yield, large energy consumption, heavy carbon emission, poor catalyst stability, and limited scalability. In this talk, I will first demonstrate a catalyst-free depolymerization method via the pyrolysis reaction that can convert model polyolefin and polyester plastics to their monomers with high yield (Nature 2023, 616, 488-494). This pyrolysis process is realized by electrified heating, which helps create a spatial temperature gradient as well as a temporal heating profile. The spatial temperature gradient promotes continuous melting, wicking, vaporization and reaction of the plastics, meanwhile the temporal heating profile by pulsed Joule heating can reduce unwanted side products and keep the reaction under far-from-equilibrium conditions. With this process and setup, we converted polypropylene and polyethylene terephthalate to their monomers via the pyrolysis reaction at record high yields of ~40%. In the second half of the talk, I will then discuss another approach to convert polyolefin plastic wastes to jet fuel and diesel precursors, also through electrified spatiotemporal heating. We demonstrate using structural effect to control the reaction progress which can allow us to achieve record-high yield to C8-C18 hydrocarbons without using any catalysts. Our preliminary TEA and LCA results indicate excellent economic potential and significant greenhouse gas emission reduction for the manufacturing process.

385l Population Balance Equations for Polymer Upcycling

Baron Peters, Ziqiu Chen, Changhae Andrew Kim, Ryan Yappert, Emmanuel Ejiogu, Jiankai GE, Lela Manis

The ongoing accumulation of plastic waste in landfills and in the environment has inspired many research efforts on polymer upcycling. To aid catalyst development and process engineering efforts, we have developed a variety of population balance models. I will outline how these models, in combination with experimental data, are being used to test mechanistic hypotheses, to extract kinetic parameters and compare catalyst activities, to assess the impacts of mass transfer, to translate ab initio calculations to observable macroscopic predictions, and to account for vapor-liquid partitioning during depolymerization in two-phase reactors. Finally, I will showcase new computational tools to automate model formulation, to solve the equations for flow or batch reactors, and to optimize parameters against experimental data.

385m Optimal Strategies of Upcycling Plastic Waste through Superstructure-Based Framework

Woochang Jeong, Jonghwan Oh, Hyeonseo IM, Yong Sang Lee, Heehyun Choi, Janggeun Lee, Jiyong Kim

The increasing environmental issue of disposing of plastic waste is a significant problem all over the world. This study focuses on analysis and evaluation of main solid-to-chemical strategies for upcycling the plastic wastes. First, we propose a technological superstructure that produces value-added chemicals from plastic waste. The rigorous simulation models of main process included in the superstructure were developed to estimate the technical, economic, and environmental parameters within the technological superstructure. Then, the superstructure along with associated parameter is used to develop an optimization model which allow us to identify and assess the optimal strategy of upcycling plastic waste under different objective functions: 1. Maximization of profit, 2. Minimization of cost, 3. Reduction of CO_2 emission, 4. Minimization of CO_2 emission, 5. Energy generation, 6. Energy consumption. The capability of the developed framework was illustrated with the production of chemicals from plastic waste and case studies from different scenarios. Based on the developed framework, we also discussed the trade-off between sorting strategy and non-sorting strategy to identify sorting performance under different scenarios. The results of this study provide applicable decision-making strategies for stakeholders and government policymakers to make decisions related to plastic waste utilization and treatment issues within the circular economy.

One of the most critical global issues is the disposal of increased plastic waste as production of plastic is faster and faster. The existing methods of disposing of plastic waste, such as landfill and incineration, have a negative effect on environment in terms of air pollution and soil contamination. Mechanical recycling is alternative technology for disposing of plastic waste, but this technology has high product yield only in high-density polyethylene (HDPE) and polyethylene terephthalate (PET). This limitation has facilitated the development of new recycling technologies and one of them is chemical upcycling. Various upcycling technologies such as pyrolysis and gasification that use plastic waste as feedstock are promising to dispose of plastic waste. However, most research focuses on the specific unit process or unit technology such as gasification of polypropylene (PP) and pyrolysis of HDPE. To overcome this limitation and to accelerate the application of upcycling technologies in industries, there are needs for evaluation of a number of practical strategies. Accordingly, the objective of this study is to develop a systematic framework for the evaluation of upcycling strategies, which can be used to answer a wide range of questions such as what plastic types would make the upcycling strategy economically viable and what types of conversion routes should be integrated and how. To develop this framework, we first generate a technological superstructure to represent and analyze a system of upcycling plastic waste consisting of various feedstocks and products along with the integration of technologies. In particular, we develop a network optimization model to identify and evaluate the optimal strategy under various objective functions. Then, we estimate the economic, environmental and energy parameters of optimization model with a rigorous process model. Finally, we identify a variety of upcycling and sorting strategies under different objective functions.

Various products such as fuel, olefin, and lubricant are produced through plastic waste depending on what type of process is used. In the technological superstructure of this study, intermediate products are generated starting with plastic waste as a raw material, and finally, a number of products are produced. To develop this superstructure, a unit process of upcycling plastic waste must be connected. We define several feedstocks, compounds, and technologies within this technology superstructure and connect the flow of materials and unit processes to complete the upcycling technology superstructure.

Based on the information on compounds and technologies in the upcycling superstructure, we developed a rigorous simulation model for each process using Aspen Plus. By utilizing the process model, we obtained the main parameters of each process. The main parameters can be classified into three categories: one is a technical parameter such as yield or energy consumption for mass and energy balance, and another one is an economic parameter related to the revenue and cost of the process. The last parameter is environmental data such as CO_2 emission or reduction. The parameters were estimated through streams of each process from the simulation model, and energy consumption was obtained through utilities used in each process. In terms of economic parameters, major parameters such as capital cost and operating cost were calculated by considering the cost and sizing of each equipment.

To present an optimal upcycling strategy using plastic waste, we develop a mixed-integer linear programming (MILP)-based optimization model. In the optimization model, there are various logical and practical constraints to meet the different objective function.

Mass and energy balance: This constraint is used to balance mass and energy flow between involved technologies from the feedstock to the final products.

Technology capacity: This constraint means that the involved technologies is limited by its capacity.

Feedstock availability: The available amount of feed that can enter a particular technology balances the amount of feed that can be accessed.

Logical constraint: In the case an input stream can be processed by various technologies, the number of technologies that can satisfy the objective function is controlled by using binary variables.

We consider various objective functions and one of them is to maximize the net profit by economic aspect. This objective function includes the terms of revenue and cost. Thus, it means that it is important to consider not only the product price but also production cost. Other objective functions that consider the environmental and energy aspect are shown in supplementary figures.

In this study, the optimal strategies for upcycling plastic waste are assessed through six different objective functions. In addition to this, we presented the optimal strategies depending on the type of plastic waste (PET, PP, poly vinyl chloride (PVC), polystyrene (PS), polyethylene (PE), mixed plastic). The composition of mixed plastic is shown in the supplementary figure. The final products of the superstructure include recycled plastics, chemicals (BTX, olefin, and aromatics), and fuels (gasoline, diesel, and lubricant), and each product is defined by carbon range and molecular structure. The availability of the feedstock is 6,250 kg/h, which is adopted from the literature. The price of feedstocks (plastic wastes) is adopted from the market price of plastic flakes which are made from pre-treatment process before feeding the upcycling system. Then, we considered the market price as the price of final products such as H_2, gasoline, diesel, and recycled PET. While the price of feedstock and product fluctuated constantly, acceptable deterministic prices were selected in this study. The utility price of each process is estimated from the utility package of Aspen Plus V12.0 simulator. The optimal strategies and specific result values under different objective functions and different types of plastic waste using the framework are depicted in the supplementary image file. In the case of maximizing net profit, PET and PVC show low economic viability due to its chemical characteristics which means that PET and PVC have oxygen and chloride, respectively. The other case of objective functions, we identified that the strategies and values from different types of plastic waste are various account of its chemical characteristics. Based on the result of each plastic waste type, we evaluate the sorting strategy when the feedstock is mixed plastic. In general, when the capacity of plastic waste increases, the sorting strategy is selected for maximizing the net profit, but when the capacity of waste increases, sorting strategy is not adopted for minimizing CO_2 emissions. The reason for this result is complexity of strategy, which means the number of processes that plastic waste goes through in the strategy. The entire results of sorting strategies are depicted in the supplementary figure.

In conclusion, this study evaluated the strategy of dealing with increment of plastic waste. In particular, to present an optimal strategy, the upcycling superstructure including several compounds such as feeds, intermediate products, and final products was developed. Based on the information of the superstructure, we estimated the technical, economic, and environmental parameters. Using these parameters and superstructure, we developed an optimization model with practical and logical constraints to meet various objective functions. Through case studies with six objectives under six different types of plastic waste, we analyzed each pathway and presented an optimal strategy for chemical production. While the developed framework is helpful in decision-making for the strategy of upcycling plastic waste, further research is needed for different scenarios. For example, the carbon tax, which is the cost of CO_2 emissions, greatly effects selecting the optimal upcycling strategy and sorting strategy.

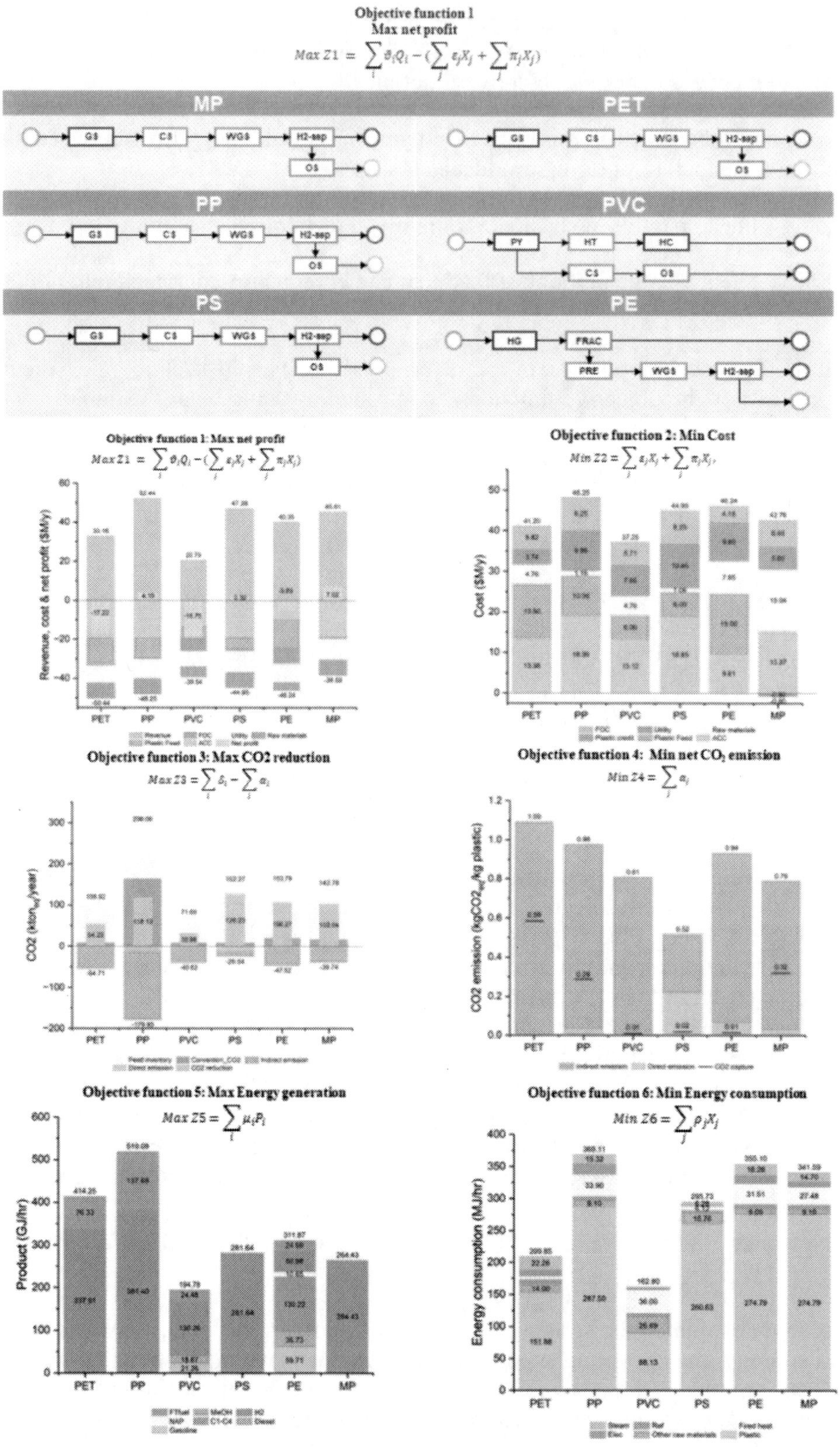

385o Recyclable Non-Isocyanate Polyurethane Network Foams from Renewable Biobased Precursors

Nathan Purwanto, Yixuan Chen, Tong Wang, John Torkelson

Polyhydroxyurethane (PHU), made by the aminolysis of cyclic carbonates, is a non-isocyanate polyurethane material with promising potential as a benign, recyclable alternative to isocyanate-based polyurethane. Recently, thiols were shown to react with cyclic carbonates to liberate CO_2, presenting a pathway to produce self-blowing PHU foams. We have developed a rheology-guided, rapid preparation of self-blowing PHU network foams, allowing for dramatic reduction of reaction times to achieve foams. Furthermore, because PHU networks exhibit covalent adaptable network (CAN) nature, the waste PHU foams synthesized by this method show excellent reprocessability to the bulk with full recovery of cross-link density and associated properties. Additionally, the use of our rheology-guided foaming strategy unlocks the potential to employ renewable biobased monomers that are less reactive, such as those derived from dimer acid and cashew nutshell liquid (CNSL). To understand further the structure-property relationships in recyclable biobased PHU network foams made with dimer-acid- and CNSL-based cyclic carbonates, we studied the effects of thiol concentration as well as functionality. We observed that morphology depends on thiol concentration but is independent of thiol functionality, whereas compressive mechanical properties depend greatly on both thiol concentration and functionality. To address further the sustainability challenges, the biobased PHU CAN foams were melt-reprocessed into bulk films with full recovery of cross-link density. We characterized the elevated-temperature stress relaxation of these bulk biobased PHU CANs, observing how the presence of permanent bonds (originating from the thiol-cyclic carbonate reaction) affects the relaxation dynamics. Moreover, the recycled bulk biobased PHU CANs function as elastomers with high dimensional stability and excellent elevated-temperature resistance to creep deformation.

385p Fermentable Organic Acids Production from Mixed Plastic Waste in the Presence of Nitric Acid

Seong-Min Cho, Sunkyu Park

Since the advent of plastics, human life without plastics has been unimaginable, and plastics are being consumed in various fields such as packaging, building construction, textiles, transportation, electronics and everyday life. Forty million tons of plastic waste was generated in the US in 2021, but the recycling rate was only ~5%, while ~85% was reclaimed in landfills and ~10% incinerated. With the concern about this unmanageable amount of plastic waste, plastic upcycling has been considered. Most accomplishments have come from polyethylene terephthalate (PET) because the polymer backbone thereof can be readily degraded with a high yield of monomers. On the other hand, although polyethylene (PE), polypropylene (PP), and polystyrene (PS) have occupied a substantial amount of plastic waste, their upcycling has been restricted because of their robust non-activated C-C linked polymer backbone. One promising approach is an oxidative deconstruction of the polymer backbone by a universal chemical process, resulting in a mixture of organic acids. This strategy can be successfully developed when the resulting mixture of organic acids is metabolized by microorganisms without purification of each organic acid, called biological funneling.

Here in, we developed a deconstruction of LDPE, HDPE, PP, and PET or mixtures thereof in the presence of nitric acid to provide fermentable organic acids. Briefly, 1 g of each plastic debris was placed in a PTFE-lined hydrothermal reactor and 60 mL of 2 M nitric acid was charged. For mixed plastic experiments, we placed 0.5 g of each LDPE, HDPE, PP, and PET, a total of 2 g of plastics. The oxidative deconstruction of plastics was conducted at 180 °C for a designated time. For PE and PP, a liquefaction of solid debris was completely achieved after 6 h, giving an aqueous mixture of acetic acid and C_{4}-$_{7}$ dicarboxylic acids, and acetic acid and lactic acid, respectively. The organic acid concentrations of the resulting solution were 10.9 g/L from HDPE, 9.6 g/L from LDPE, and 10.2 g/L from PP. For PET, the vanishment of solid debris was completely achieved after 8 h, giving solid crystal of terephthalic acid with a yield of 89.5 mol%. When all 4 different plastics were mixed together, the resulting solution after 8 h of deconstruction had a total of 20.0 g/L of acetic acid, lactic acid and C_{4}-$_{7}$ dicarboxylic acids, and a comparable yield of terephthalic acid was obtained. The aqueous organic acid solutions and terephthalic acid were "ready for feeding" to microorganisms after pH adjustment without further purification.

385q Biocomposite Thermoplastic Polyurethanes Containing Evolved Bacterial Spores As Living Fillers to Facilitate Polymer Disintegration

Adam Feist

The field of engineered living materials (ELMs) seeks to pair living organisms with synthetic materials to generate biocomposite materials with augmented function since living systems can provide highly-programmable and complex behavior. ELMs have typically been fabricated using techniques in benign aqueous environments, limiting their application. In this work, biocomposite fabrication was demonstrated in which spores from polymer-degrading bacteria, *Bacillus subtilis,* were incorporated into a thermoplastic polyurethane (TPU) using high-temperature melt processing. Bacteria were engineered using adaptive laboratory evolution to improve their heat tolerance to ensure nearly complete cell survivability during manufacturing at 135° C. Mutation analysis of the multiple replicate evolutions identified the genetic basis for enhanced heat-shock tolerance of the Bacillus spores. Furthermore, the overall tensile properties of spore-filled TPUs were substantially improved, resulting in a significant improvement in toughness. The biocomposites facilitated disintegration in compost in the absence of a microbe-rich environment. Finally, spores retained a programmed function, expressing green fluorescent protein. This research provides a scalable method to fabricate advanced biocomposite materials in industrially-compatible processes.

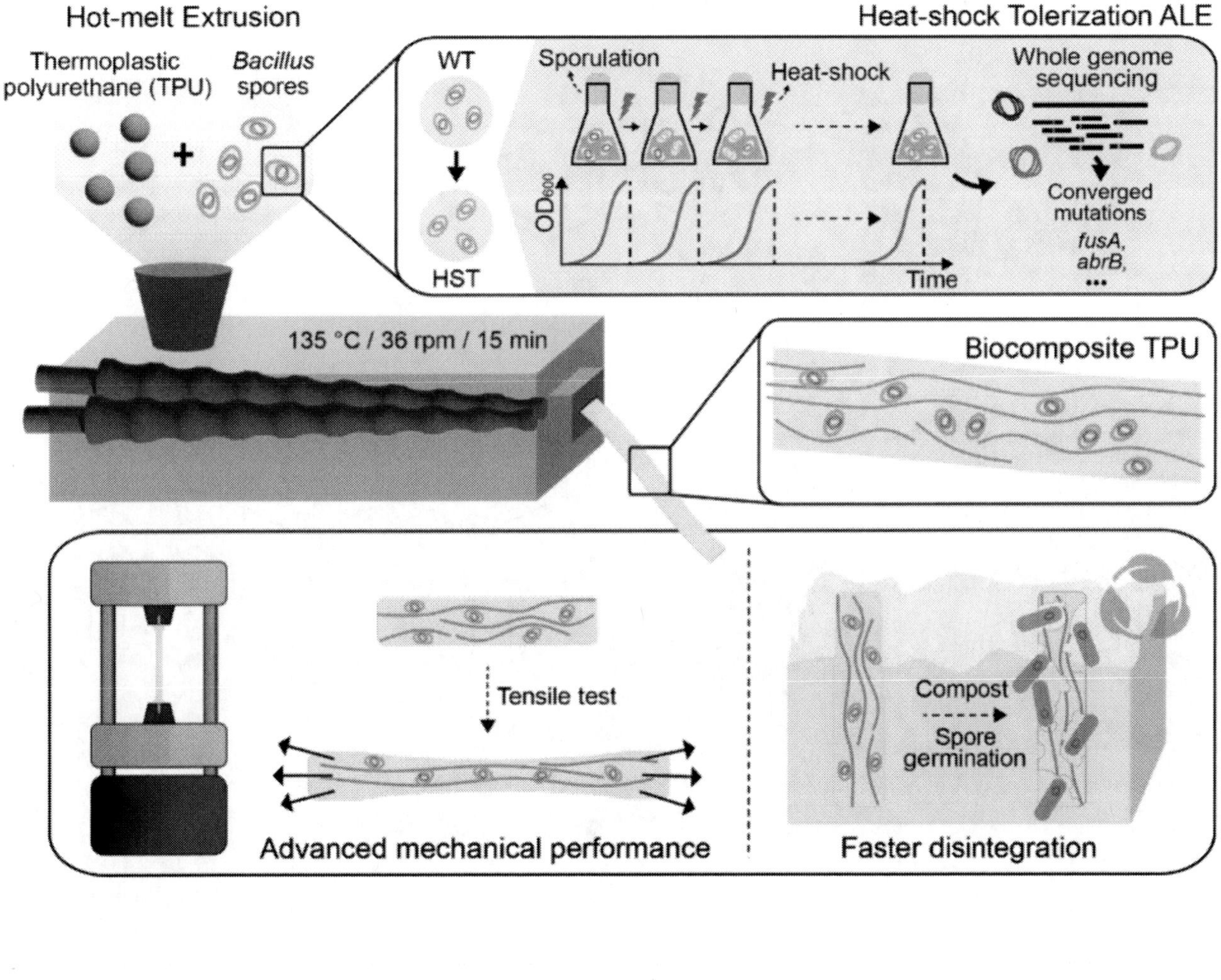

385s Post-Polymerization Functionalization of Waste Poly(styrene) Using Ionic Liquids Chemistries

Zahra Sekhavat Pour, Jason Bara

The global increase in poly(styrene) (PS) manufacturing has led to significant environmental impacts due to its widespread use and accumulation in landfills. Traditional mechanical recycling of PS encounters challenges due to its chemical structure, low bulk density, and limitations in sorting techniques. These issues often result in the production of downcycled materials with reduced quality and utility. Modification of commodity plastics through C-H functionalization methods is a growing focus in polymer science. This approach has potential applications in upcycling plastic waste sustainably and creating new functional materials with unique properties not easily obtainable through traditional synthetic routes. In our study, C-H functionalization of waste PS was conducted using Friedel-Crafts acylation, followed by subsequent modification using imidazolium-based ionic liquids. The resulting ionomers contained methyl, butyl, or hexyl imidazolium cations bound to the PS backbone and "free" bistriflimide anions (Tf_2N^-). The materials obtained demonstrated unique characteristics compared to the waste PS. Particularly noteworthy was the significant reduction in the glass transition temperature (T_g) of the butyl imidazolium-functionalized PS, decreasing by almost 100 °C. This transformation turned the initially brittle PS into highly stretchable materials comparable to elastomers. These ionomers create possibilities for exploring the use of plastic waste as a potential source for producing advanced materials.

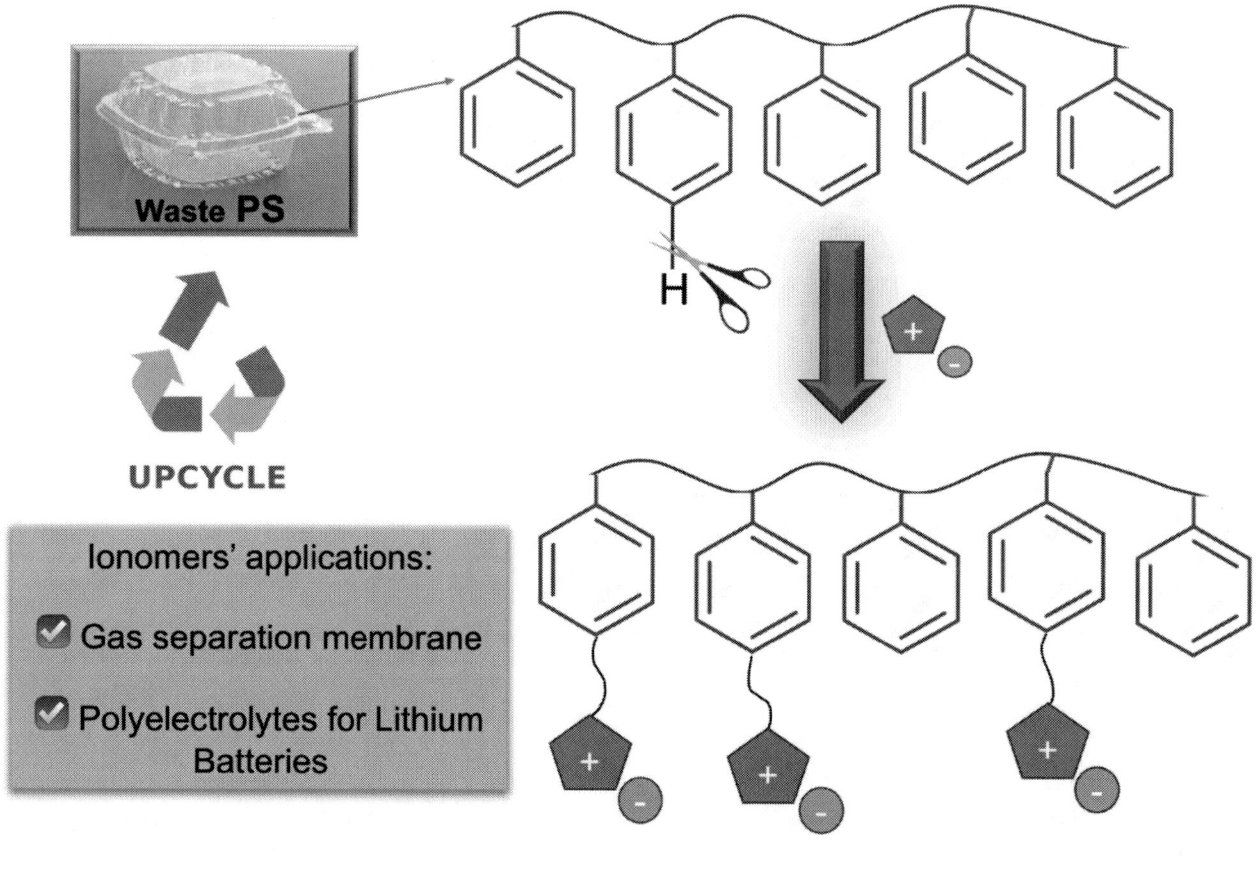

385t Making Non-Recyclable Polymers Recyclable: Novel Advances in Obtaining Value, Circularity, and Sustainability from Spent Polymer Networks and Thermoplastics

John Torkelson

Here, we will discuss recent advances from our research group regarding circularity, including chemical recycling and small-molecule recovery, from covalent adaptable networks (CANs) that can serve as recyclable replacements for non-recyclable polymer thermosets with permanent covalent cross-links. Examples will be drawn from two classes of materials.

The first class is networks from addition-type polymers, e.g., polyethylene (PE), ethylene-based copolymers (EBCs), and polymethacrylates. We will discuss the use of a dynamic covalent cross-linker which can be to develop CANs either by free-radical copolymerization at low levels with other monomers or by radical-based, melt-state reactive processing, e.g., extrusion, with thermoplastics such as PE, EBCs, or polymethacrylates with hexyl (e.g., poly(n-hexyl methacrylate) or longer side groups. Such networks can be recycled (reprocessed) multiple times with full recovery of cross-link density and associated properties. Additionally, the radical-based reactive processing approach allows for spent or waste thermoplastic PE, EBCs, etc. to be upcycled into higher value, reprocessable networks with enhanced properties.

The second class is condensation-type networks from biobased non-isocyanate polythiourethanes (NIPTUs), which can serve as benign replacements for conventional polyurethane networks made from toxic isocyanates. This is important because polyurethane is the sixth most produced polymer, with the vast majority being in network form. NIPTU CANs exhibit full recovery of cross-link density and associated properties after multiple reprocessing steps, providing an excellent route to recycle spent materials into valuable products. Additionally, NITPU CANs exhibit rapid synthesis and reprocessability due to the "click"-like nature of the chemistry. We have also demonstrated that robust properties can be obtained with NIPTU networks made from biowaste-based starting materials, e.g., cashew nutshell liquid and glycerol, the latter being a waste byproduct of biodiesel production. Additionally, we will show that end-of-life NIPTU CANs may allow for facile recovery of valuable small molecules by chemical recycling, with a specific example providing for 97% yield of recovered small molecule.

385u Nutrient Sourcing and Process Optimization in Microbial Plastic Waste Valorization and a Molecular Level Study of the Effects of Nutrients of Target By-Products

Kimia Noroozi, Laura R. Jarboe

Plastic waste has been a long-standing environmental and societal issue in the modern world, and the subject of plastic recycling is extensively studied. Biological upcycling of plastic waste is an attractive topic and the prospects of using microbial cell factories include sustainability, circular economy, and a potential for production of value-added products. While a variety of organisms have been studied for their ability to utilize plastic waste, the complexity and stability of molecules present in plastic extend the process time and decrease its yield. The oleaginous yeast *Candida maltosa* is one of the species that has been studied for its ability to utilize representative hydrocarbon molecules. While *C. maltosa*'s impressive ability to grow on hydrophobic substrates as carbon sources introduces the possibility of biological upcycling plastic waste, the efficiency of the process needs to be targeted to make it a viable option for scale-up. To this end, it is essential to look into media components and their effects in the uptake process, while strategizing nutrient sourcing to enhance performance by replacing conventional media components with those that have a more pronounced impact on the yield. This process can also help in identifying potential product targets for the valorization of plastic waste into value-added products. Looking at the molecular impacts of nutrient optimization can provide valuable insights into the use of microbial systems for plastic upcycling.

Results and Conclusions

In this study, various organic and inorganic nitrogen sources including ammonium sulfate, urea, and casamino acids were tested to determine their effects on biomass production when hydrocarbon molecules are used as the sole carbon source in the media. Yeast cells are equipped with a nitrogen catabolic repression system that senses the type of nitrogenous compounds present in the environment and use them according to cell selectivity. Ammonium sulfate is expected that it would result in the highest growth possible. Urea, which is an organic nitrogen source and the main component of urine can be used to replace ammonium sulfate in this process without sacrificing the yield. In addition to supporting growth, urine is a waste stream that can be effectively utilized in this process.

Furthermore, when hydrocarbon-based substrates were used in conjunction with replacing ammonium sulfate in the media with casamino acids, a general trend of increase in final biomass production was observed, which is especially important for longer-chain hydrocarbon compounds. Hence, a replacement of the nitrogen source is considered an effective way to increase the uptake. The use of casamino acids as the sole nitrogen source for the process was shown to also affect intracellular protein accumulation in most cases, compared to when glucose was used as the carbon source. Proteomics data shows that proteins involved in fatty acid production pathways are found in the cells grown in hydrophobic substrates which can point to a valuable set of byproducts that can be targeted for valorization.

Aside from this, phosphorus and other components of the media were also studied for replacement and optimization. Different phosphorous compounds were also tested at various concentrations and proved not to have an effect. Also, all the vitamins and trace metals in standard media were replaced by a human multivitamin which resulted in an increased growth, while significantly reducing process costs. Process conditions such as aeration also showed a significant effect in increasing biomass production.

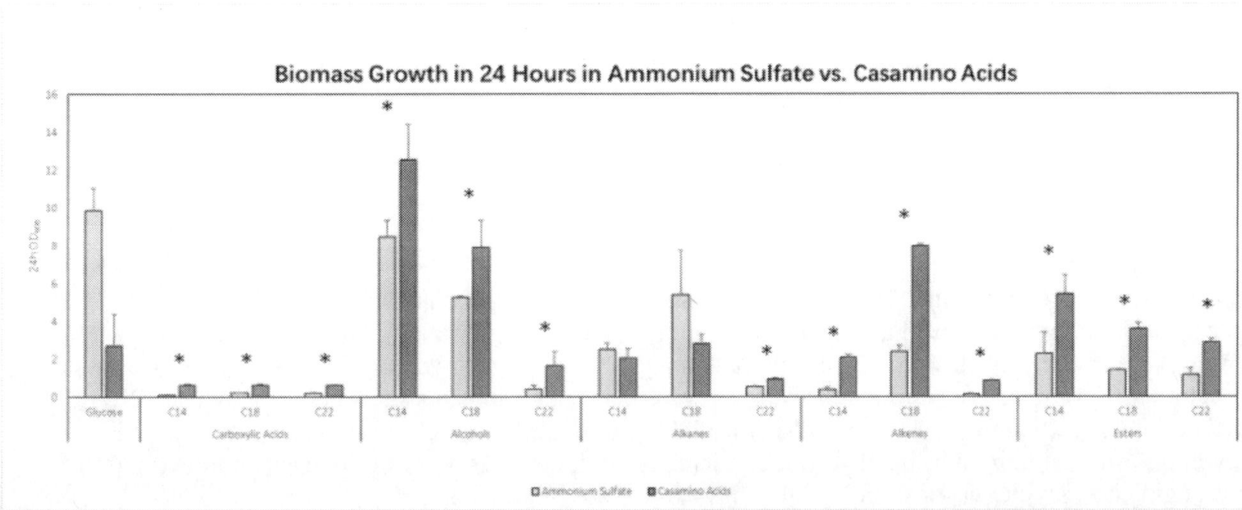

Figure 1 – 24-hour growth of *Candida maltosa* on model hydrocarbon compounds of different functional groups with 14, 18, and 22 carbon chain lengths grown in YNB with Ammonium sulfate or Casamino acids as the nitrogen source. Stars (*) above columns indicate significantly higher growth of casamino acid cultures compared to ammonium sulfate.

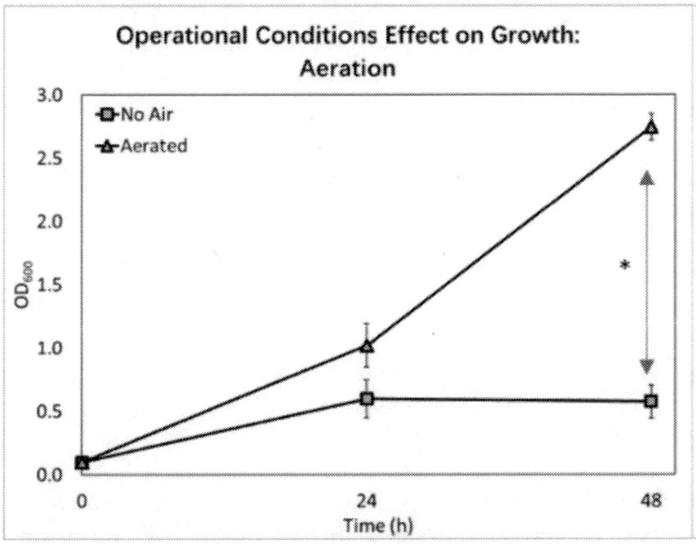

Figure 2 – Growth curve of *Candida maltosa* grown on thermally oxo-degraded plastic waste in an aerated and non-aerated fermenter.

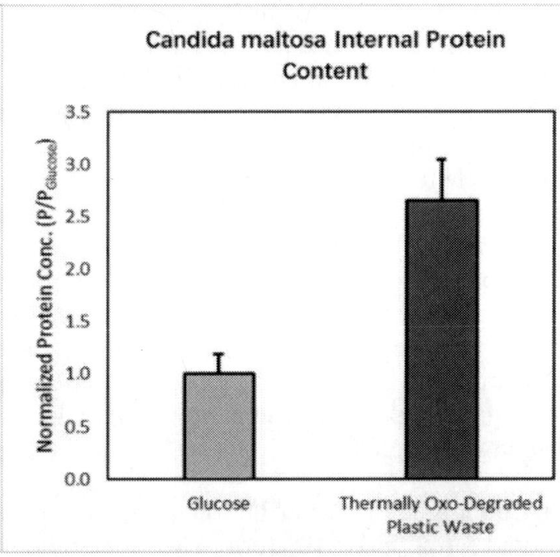

Figure 3 – Normalized protein content of *Candida maltosa* grown on glucose and thermally oxi-degraded plastic waste.

385v Reactive Molecular Dynamics-Guided Deconstruction of Single-Use Plastics for Value-Added Chemicals and Novel Materials

Tridip Das, William Goddard III

The growing environmental concern surrounding single-use plastics necessitates innovative approaches for their sustainable management. In this study, we investigate the catalytic deconstruction of plasma-treated single-use plastics, aiming to transform them into value-added chemicals and novel materials. Leveraging molecular dynamics simulations, we elucidate the underlying mechanisms of plastic breakdown at the atomic level. Our findings provide crucial insights into designing efficient catalysts, considering reaction conditions, and predicting the formation of novel materials during the deconstruction process. This interdisciplinary approach bridges materials science, catalysis engineering, and computational modeling, offering a promising avenue for addressing plastic waste challenges. Among different single-use plastics, polyethylene terephthalate (PET) is the most recycled, but other plastics such as polyethylene (PE), polypropylene (PP), and polystyrene (PS) contribute significantly to pollution due to difficulties in recycling, often ending up in landfills or aquatic environments. Our research focuses on upcycling these challenging-to-recycle plastics like PE, PP, and PS.

In our research, we investigate the breakdown mechanism of polymer chains using ab initio molecular dynamics. Subsequently, we employ reactive molecular dynamics (RMD) to predict all possible products under varying reaction conditions. Specifically, we simulate the plasma treatment process, considering potential free radical reactions. For instance, oxygen plasma can generate ozone as a reactive species. This ozone then reacts with hydrocarbons in plastics, leading to the formation of additional radicals such as H*, OH*, and HO2*. Our ab-initio and RMD simulations provide evidence for these processes. We conducted Ab Initio Molecular Dynamics (AIMD) simulations using the Vienna Ab Initio Simulation Package (VASP) to obtain accurate electronic structure information. These simulations account for quantum mechanical effects, providing insights into the behavior of polypropylene breakdown and product formation at the atomic level. From AIMD trajectories, we extract local configurations around polypropylene molecules, which serve as training data for the ReaxFF force field. By interpolating between known AIMD results, we learn the force field parameters, significantly reducing computational costs compared to fully quantum-mechanical simulations. To validate the ReaxFF, we compare its predictions with additional AIMD simulations on plasma-treated plastics in water medium. The force field's accuracy is assessed through thermodynamic properties, vibrational spectra, and structural features. Finally, we successfully compare the predicted results with experimental observations. Beyond deconstruction, the ReaxFF enables predictions of novel materials that may form during the breakdown of plasma-treated polypropylene, providing valuable insights for designing sustainable alternatives and recycling strategies.

In a groundbreaking contribution to the literature, we present the **-C-C-** polymer breakdown mechanism captured through **AIMD** study. The process begins with an activated ozone radical (O-atom) derived from an ozone molecule. Surprisingly, despite its proximity to a primary hydrogen (H), the O-atom targets the secondary hydrogen (secondary-H), which possesses a **C-H bond 4 kcal/mole weaker**. Through a multistep sequence, this H-atom is transferred to another -O radical, resulting in the formation of an -OH radical. Subsequently, the -O radical attaches to the secondary carbon, leading to the formation of a **-C=O bond**, ultimately breaking the **C-C backbone**. The other end of the broken iPP chain has its H-atom react with a free -O radical to form H2C=CH-. The breakdown process leads to chain scission, resulting in a reduction in the molecular weight of the polymer. Smaller polymer fragments are more susceptible to further degradation. During the breakdown process, reactive species (such as H*, OH*, and HO_2* radicals) are generated. These species can further participate in chemical reactions, leading to the eventual degradation of the plastic. We report here the formation of C2-C4 chains and oxygenated intermediates like ethers and ketones along with HCHO and HCOOH, but these products were dominated by small gas phase products such as CH_4, H_2O, H_2, CO, CO_2. In summary, understanding polymer breakdown and small molecule formation mechanisms guides sustainable alternatives and recycling strategies, contributing to a circular economy.

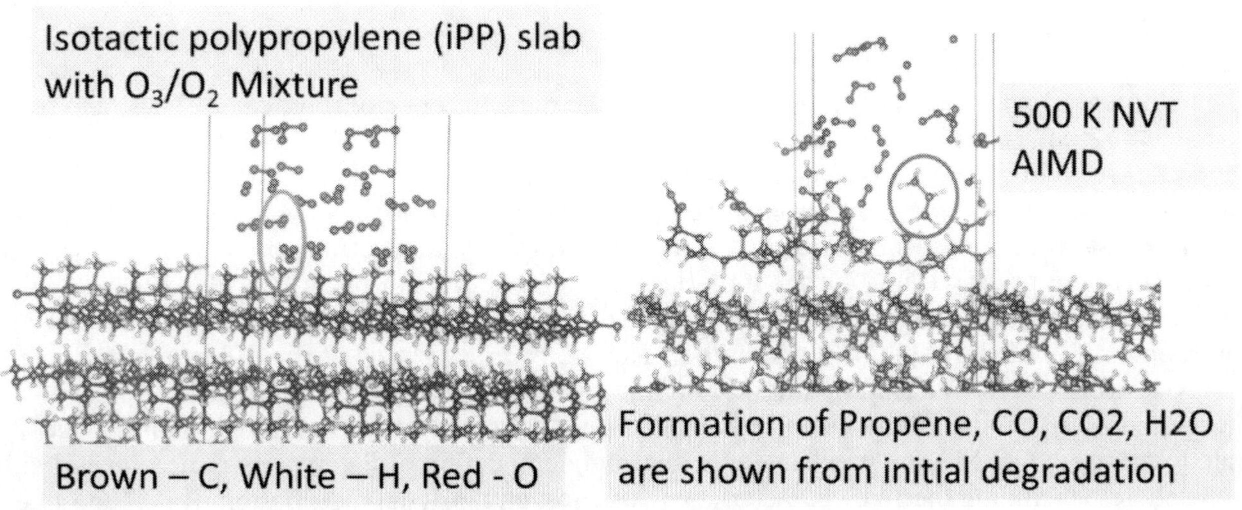

385w Two-Stage Pyrolysis of Polyethylene to Light Olefins Assisted By CO_2

Qidian Sun, Yi Cheng

The treatment of waste plastics is currently one of the most urgent environmental issues, and reducing carbon emissions is also a widely recognized focal point. Pyrolysis to obtain high-value products is a method of recycling waste plastics that has garnered significant attention from researchers. Inert gases such as nitrogen are the most commonly used atmospheres in waste plastic pyrolysis studies. Some researches have utilized carbon dioxide for dry reforming to obtain carbon monoxide and hydrogen gas following pyrolysis. Using carbon dioxide as the atmosphere for plastic pyrolysis not only allows for the recycling of waste plastics but also reduces carbon dioxide emissions, making it environmentally friendly. In this work, carbon dioxide is directly used as the PE (polyethylene) pyrolysis atmosphere. We employ thermogravimetric analysis (TGA) combined with infrared spectroscopy (IR) and mass spectrometry (MS), as well as pyrolysis/gas chromatography/mass spectrometry (PyGC-MS), to investigate the thermal degradation of PE in carbon dioxide and nitrogen atmospheres, including thermal characteristics, kinetics, and product distribution. It is found that there is almost no difference in the thermal degradation curves and gas product yields of PE in the two atmospheres at low temperatures, indicating that carbon dioxide in the pyrolysis process of PE was not activated. Further experiments are conducted on the pyrolysis of PE at temperatures exceeding 1000 °C in a two-stage process. At such high temperatures, carbon monoxide is detected in the products in CO_2 atmosphere, demonstrating the activation of carbon dioxide in the reaction. By manipulating the reaction time, high-value products such as light olefins can be obtained. A hydrogen-rich gas is added to balance the C/H ratio in an attempt to obtain more light olefins.

385x Acid-Catalyzed Decomposition of Polypropylene into Naphtha in Hydrocarbon Solvent

Masahiko Matsukata, Mahiro Matsushita, Tohru Kamo, Motomu Sakai

Focusing on the chemical recycling of waste plastics, especially polyolefins such as polyethylene and polypropylene, we aim to develop a catalytic conversion process of waste plastics into petrochemical feedstock. The challenges in chemical recycling of waste plastics are the extremely high viscosity of melted plastic, which makes it difficult to handle as a fluid in the process, the associated poor heat conduction and mass transfer in the reactor, and the need to remove various foreign substances such as fillers. In addition, it is generally known that pyrolysis produces a large amount of light oil components. To address these issues, we propose the use of hydrocarbon solvents to improve flowability as a fluid, removal of solid foreign substances, and conversion to petrochemical feedstock (C5-C9 products) by catalytic cracking using zeolites, assuming that the petroleum refining process would be utilized.

In this study, polypropylene (PP) was used as model plastic and zeolite beta as the catalyst. Hydrotreated light gas oil (RBC-LGO) and light cycle oil (RBC-LCO) were used as petroleum solvents. RBC-LGO had a higher content of aliphatic hydrocarbons, while RBC-LCO had a higher content of aromatic hydrocarbons. In addition, n-cetane (n-C16) and 1-methylnaphthalene (1-MN) were used as solvents representing linear alkanes and aromatic hydrocarbons. The decomposition behavior of PP in these petroleum solvents, e.g., PP conversion, was studied.

Decomposition test was conducted in a batch reactor at 673 K for 60 min using 20 g of solvent, 5 g of PP, and 1 g of beta zeolite. As shouwn Figure 1, while the levels of PP conversion was 97.0% and 58.4% when RBC-LGO and RBC-LCO were used as solvents, those were 96.9% and 68.2% when n-C16 and 1-MN were used as solvents, suggesting that aromatic hydrocarbons had a negative effect on the PP decomposition because the high aromatic hydrocarbon content in LCO inhibited the reaction of aliphatic hydrocarbons within the micropores of zeolite beta. The yields of C2-C9 petrochemical feedstock were 60-70%, depending on the reaction conditions, indicating that zeolite is a suitable catalyst for producing petrochemical feedstock. As a comparison, amorphous silica-alumina was also used as a catalyst, but the yield of petrochemical feedstock was low. Therefore, it was found that aliphatic hydrocarbon-rich solvents such as LGO are suitable for the cracking of PP.

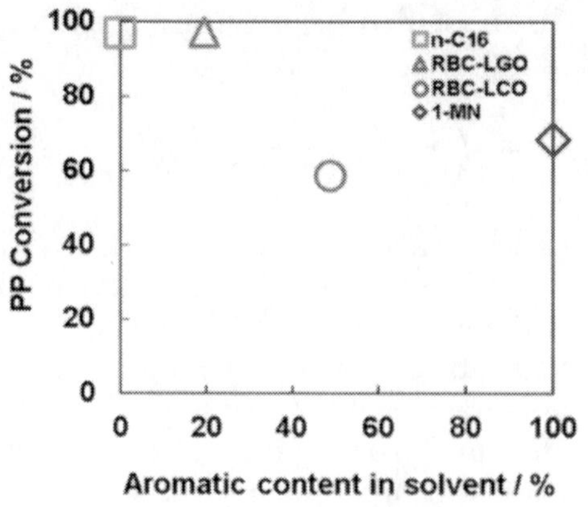

Figure 1 Relationship between aromatic content in solvent and PP conversion after 60 min at 673 K.

385y Understanding the Pyrolysis Process of Polyethylene By a Coupled Method of Machine Learning and Molecular-Level Kinetic Model

Maoxian WANG, Yi Cheng

The industrialization and the development of the global economy have promoted the production of plastics, resulting in an overwhelming amount of waste plastic that requires effective solutions for resource utilization. Pyrolysis can convert plastic into higher value-added gas/liquid products, which can be utilized as raw materials in the chemical industry.

Several studies have been conducted in experimental aspect of waste plastic pyrolysis, but it is difficult to obtain a comprehensive understanding of the process from scattered experimental data due to its complexity. Thus we adopted the structure-oriented lumping (SOL) method to provide an approach of a molecular-level kinetic model that effectively utilizes limited pyrolysis data. Based on PE pyrolysis system, the model automatically generates complex reaction networks, realizing rapid analysis of data and optimization of model parameters. It also retains the basic information of the reaction mechanism, being suitable for rapid simulation and production guidance of pyrolysis designs.

Machine learning algorithms with powerful data fitting ability can assist in the accurate prediction of product yields and thus guide the scale-up design of pyrolysis processes, but they often face challenges due to insufficient experimental data samples and large search spaces. We extrapolate the molecular weight distribution of the raw materials to obtain sufficient data, which is then re-entered into the molecular-level kinetic model to compute the distribution of gas and liquid phase products. Subsequently, we use the raw material distribution, system parameters, condition parameters, and kinetic parameters as input factors, with gas and liquid yields as output targets, to construct machine learning models and achieve excellent predictive accuracy. Additionally, partial dependence plots quantify the impact among input factors and their effects on yield, providing in-depth insights into the pyrolysis process and enhancing yield optimization.

In conclusion, this study not only advances the theoretical framework for polymer degradation but also provides practical insights into the optimization of pyrolysis processes, offering a pathway to efficient waste plastic recycling and the sustainable production of valuable chemicals.

385aa Upcycling Linear Low-Density Polyethylene Waste into Graphene for High Mass Loading Supercapacitors

Yuan Gao, Ngoc Tien Huynh, Ki-Joong Kim, Congjun Wang, Christopher S. Matranga, Viet Hung Pham

Plastic pollution is triggering a global environmental crisis, prompting research efforts in recycling and upcycling to address this challenge. Upcycling plastic into value-added carbons, such as graphene and porous carbon, offers attractive options to manage waste streams by converting the plastic into carbon electrode materials for energy storage devices. Polyethylene is notoriously difficult to upcycle because it decomposes into light gases at approximately 350-400 °C which prevents processing it at higher temperatures to convert it into electrode materials. We address this challenge by using an air-based, thermo-oxidative process along with a solid surface area enhancer, which heats linear low-density polyethylene (LLDPE) just below the decomposition point to initiate oxidation and cross-linking of LLDPE alkyl chains. These molecular transformations allow LLDPE to be processed into char at approximately 800 °C or graphenized at 950 °C without decomposing. The LLDPE-derived graphene (LLDPE-G) has a Brunauer–Emmett–Teller (BET) specific surface area up to 1,800 m^2/g and Raman I_D/I_G ratio of 0.85. When used as electrode material for a symmetric supercapacitor with the 1 M H_2SO_4 electrolyte, LLDPE-G possesses an outstanding specific capacitance up to 175 F/g at mass loading of 20 mg/cm^2, twice the commercial requirement, yielding an excellent areal capacitance of 3.5 F/cm^2. Moreover, LLDPE-G exhibits exceptional cycling stability with capacitance retention of 95.8% after 100,000 cycles at high current density of 4.0 A/g. Last, but not least, KCl is recycled and reused over three cycles with material quality and electrocapacitive performance of LLDPE-G retained and verified after each cycle. Our approach creates new opportunities for upcycling not only waste LLDPE but also other PEs to high value graphene materials for electrochemical energy storage device applications.

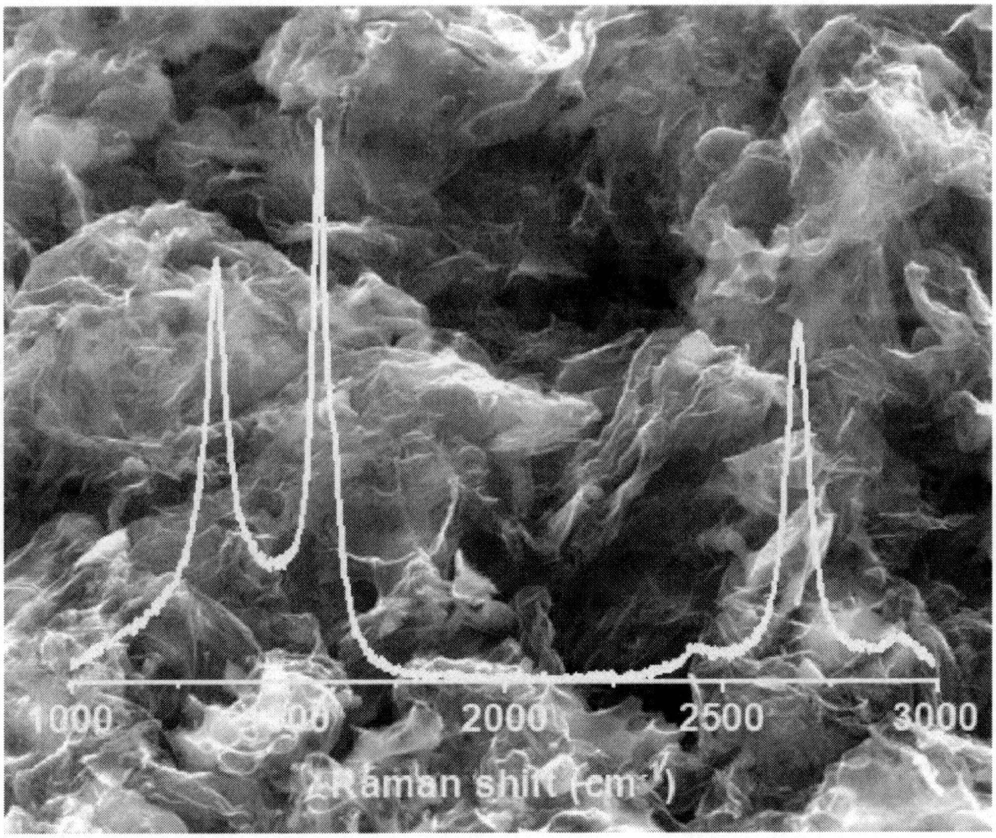
1000
2000
2500
3000
Raman shift (cm⁻¹)

385ac Effects of Capping Agents on Deblocking Temperatures of MDI-Based Blocked Isocyanates: A Study of Nucleophilicity, Pka, and Chain Length Trends

Remsha Rafiq, Jessica Anne Gondak, Hannah Zucco, Stacey H Ngo, Daylan Sheppard, Joeri Plusnin, Marie-Anne Persoons, Glenn Cormack, Goetz Veser

Polyurethanes are the 6th most used synthetic polymer in the world with an annual production of 26 MMT in the year 2022. Thermoplastic polyurethanes (TPU) are a class of polyurethanes that distinguishes itself by its capacity to be melted and reshaped, unlike thermoset polyurethanes. The widespread use of TPU has resulted in substantial waste in landfills and oceans, necessitating effective recycling solutions to address environmental concerns. Chemical recycling of TPUs presents a sustainable approach towards circular re-use of waste polyurethanes.

The present work focuses on utilizing blocked isocyanate chemistry as a strategic approach for TPU depolymerization, involving the capping of reactive isocyanate groups during depolymerization in order to passivate them for separation, followed by thermal deblocking to regain their reactivity for repolymerization within a four-step chemical recycling process that involves 1) catalytic depolymerization and capping of TPU, 2) separation of the capped hard and soft segments, 3) thermal dissociation to recover the capping agent and (uncapped) hard segment segments, and 4) repolymerization of fresh TPU from the recovered hard and soft segments. In order to make this scheme feasible, efficient, and economic, the selection of capping agents is critical. A suitable capping agent needs to deblock at low temperature to minimize the energy intensity of the process while remaining stable at the temperature of the depolymerization step. Understanding the deblocking temperatures of different capping agents is hence critical. Based on a thorough literature review of potential capping agents, we synthesized capped 4,4'-methylene diphenyl diisocyanate (MDI) using select capping agents, determined deblocking temperatures of these model compounds using thermogravimetric analysis (TGA), and derived correlations between deblocking temperature and physicochemical properties of the capping agents, such as nucleophilicity, pKa, and chain length. The results reveal clear trends, with specific functional groups and molecular characteristics affecting the deblocking temperature. The results hence yield a guideline for the molecular design of effective capping agents for blocked isocyanate chemistry, enhancing the feasibility of TPU depolymerization and promoting circular economy practices in polymer recycling.

385ad Electrochemically Enhanced Hydrolysis and Upcycling of Polyethylene Terephthalate (PET)

Zenifar Haque, Dayana Donneys Victoria, Gerardine Botte

The broad commercial use of polyethylene terephthalate (PET) is due to its attractive physical and chemical properties, including high transparency, crystallization rate, high thermal stability, and high mechanical properties, leading to a massive annual production of 70 million tons—making up 10% of all plastics [1]. However, these robust physicochemical properties of PET contribute to slow degradation in nature and the formation of environmentally damaging microplastics, which pose risks to the environment and human health. Therefore, emphasizing the necessity for its recycling. Besides mechanical recycling methods, extensive study has been conducted on chemical hydrolysis in breaking down PET into its monomers (ethylene glycol and terephthalic acid). Recently, electrochemical methods have gained attention as a promising alternative to convert PET hydrolysate (ethylene glycol) into valuable chemicals [2].

Botte's group is investigating electrochemical approaches to convert polyethylene terephthalate into valuable chemicals such as formic acid, formate etc. This novel approach integrates the hydrolysis of PET with the concurrent electrochemical oxidation of ethylene glycol. Our studies indicate that the electrochemical approach enhanced PET hydrolysis in comparison to traditional chemical hydrolysis [3] along with the oxidation of ethylene glycol into formic acid/formate in a single process, eliminating the need for separate hydrolysis. In this presentation, we will cover the outcomes of enhancing the hydrolysis of PET through electrochemical means, employing transition metals as catalysts, and using a low potential under mild temperature conditions. Optimization of this process could be a sustainable approach for upcycling of PET into valuable products.

1.Liu, K., et al., *Selective electrocatalytic reforming of PET-derived ethylene glycol to formate with a Faraday efficiency of 93.2% at industrial-level current densities.* Chemical Engineering Journal, 2023. **473**: p. 145292.

2.Li, Y., et al., *Alcohol–alkali hydrolysis for high-throughput PET waste electroreforming-assisted green hydrogen generation.* Journal of Materials Chemistry A, 2024. **12**(4): p. 2121-2128.

3.Lu, F. and G.G. Botte, *Understanding the Electrochemically Induced Conversion of Urea to Ammonia Using Nickel Based Catalysts.* Electrochimica Acta, 2017. **246**: p. 564-571.

385ae Hydrothermal Clothing Recycling

Aristidis Mihalos, Alex Paulsen, Andrew Wagner

About 17 million tons of clothing is produced every year in the U.S. and 66% ends up in landfills. Natural cotton and wool eventually break down to produce leachates that can enter the water table and methane that is a major greenhouse gas. Synthetic polyester, nylon, spandex, and acrylic clothing persists as an environmental pollutant that disintegrates into microplastics that have been found in human bodies and even freshly fallen Antarctic snow. Clothing is exceptionally difficult to recycle because most clothes are a mixture of natural and synthetic fabrics that cannot be manually separated. As a result, most reused clothing is downcycled to fillers and insulation. In this work, we examine the feasibility of using a catalytic, hydrothermal process to react cotton to water soluble products and recover solid polyester. With an acidic catalyst, the cotton undergoes hydrolysis to form soluble sugars and sugar-derivatives. The polyester is unreacted and can be recycled into new clothes and other products. We tested conditions to maximize polyester recovery and cotton conversion. We will discuss the effect of different reaction conditions (temperature, residence time, catalyst, pre-treatment, loading) and the resulting product purity, breakdown, and yield.

385ag CO_2 Induced Separation of Terephthalic Acid for Polyester Upcycling

Diego Trevisan Melfi, Aaron M. Scurto

Alkali decomposition of polyesters for polymer upcycling is a well-established depolymerization pathway. Particularly, decomposition of polyethylene terephthalate (PET) with sodium hydroxide has been shown to be highly productive and selective towards disodium terephthalate (Na_2TP). Traditionally, strong mineral acids (e.g. HCl or H_2SO_4) are then used to protonate and precipitate the original monomer, terephthalic acid (H_2TP) that is virtually insoluble in water, from aqueous disodium terephthalate solutions. As a potential sustainable alternative to strong acids, we are investigating the use of gaseous carbon dioxide (CO_2) as an acid switch for the recovery of terephthalic acid from aqueous disodium terephthalate solutions.

In aqueous solution, disodium terephthalate is generally dissociated as sodium (Na^+) and terephthalate ions (TP^{-2}). Upon the addition of CO_2, simultaneous phase and chemical equilibrium will occur. Gaseous CO_2 dissolves as aqueous CO_2 and simultaneously hydrates (carbonic acid) and dissociates to carbonate ions (HCO_3^- and/or CO_3^{-2}) and aqueous protons. From a solution of Na^+, H^+, terephthalate, bicarbonate and/or carbonate ions, terephthalic acid precipitation occurs.

The identity and quality of the produced terephthalic acid was verified by FTIR, NMR, ICP, and optical density measurements. Terephthalic acid yields up to 94 % were verified experimentally. Low Na_2TP loadings, low temperatures, and high pressures favor the recovery of terephthalic acid. The amount of terephthalic acid that can be recovered from aqueous disodium terephthalate solutions seems to be limited in some part by a buffering effect caused by the present of the co-product, sodium bicarbonate. Modeling using equations of state and activity coefficient models is presented.

385ai Surface Modification of Polypropylene Using Dielectric Barrier Discharge (DBD) Plasma for Polymer Upcycling

Shashwata Chakraborty, Sujoy Bepari, Debarati Mukherjee, Debasish Kuila

Plastic waste has become a major threat to environment due to contamination of soil, air, and oceans. We are interested in plastics upcycling of polyolefins via surface functionalization, followed by catalytic deconstruction to olefins. The current study addresses the effect of surface functionalization of polypropylene (PP) using non-thermal plasma (NTP) at atmospheric pressure. This approach to tailor the surface properties of PP is promising as it can influence the kinetics of chemical conversion reactions. The PP beads were treated with DBD air plasma at varying frequencies- 330, 565 and 800 Hz for 30 mins. The surface of NTP treated PP was investigated with respect to time for 40 days using different characterization techniques at different time intervals. The FTIR studies confirmed the presence of different functional groups such as C=O, C=C and N-O on the surface and they are consistent with the XPS analysis. In addition, distinct peaks of C-C/C-H and O-C=O are observed in the XPS spectra. However, the intensity of the peaks decreased with time. The FESEM study shows that the polymer beads disintegrated to form small particles with time. The details of these studies will be presented.

385aj A Comprehensive Study of the Impact of Polyolefin Structure on Pyrolysis Plastic Oil Composition

Jiayang Wu, Zhen Jiang, Manos Mavrikakis, George W Huber

Seven types of plastics with differing sources, molecular weights, and polymer structures underwent pyrolysis in a fluidized bed reactor: PCR HDPE, PCR PP, virgin resins of varying molecular weights of HDPE, virgin resins of LDPE, LLDPE, and PP. Each feedstock's structure was characterized using FTIR, GPC, and NMR. The pyrolysis process yielded non-condensable gases (C1-C4), oil phase products (C5-C40), and solids (C40+ and chars). The predominant compounds found in the pyrolysis oils were alkane, alkene, alkadiene, aromatic, and multicyclo-aromatics, with variations in composition observed across different plastic feedstocks.

Among the three factors examined, polymer structure exerted the greatest influence on pyrolysis product distributions, while molecular weight had the least impact. Branches in PE were identified as thermal defects where polymer degradation initiated. DFT calculations revealed that, in the absence of branches or with a methyl group branch, the cleavage of the polymer C-C backbone in HDPE required 365 kJ/mol. Conversely, in polymers like LDPE and LLDPE, where branches contained more than one carbon, branch cleavage from the backbone occurred first, requiring lower energy (305 kJ/mol). Higher branch density in PE resulted in increased concentrations of aromatics, branched alkanes, and alkanes due to more occurrence of secondary reactions. However, PP and PE exhibited distinct degradation mechanisms, with PP requiring less energy for decomposition and producing more pyrolysis oil than PE under equivalent conditions.

Pyrolysis oil derived from PCR HDPE and PCR PP contained a higher proportion of branched compounds compared to virgin plastics. Impurities, such as trace elements present in PCR plastics, could potentially promote the isomerization of linear hydrocarbons to branched forms during pyrolysis. The presence of sand acted as a getter during pyrolysis to remove these impurities. While pyrolysis effectively eliminated most trace elements, additional purification steps may be necessary for the resultant oils.

385ak Catalytic Dehydration of Polyvinyl Alcohol over Solid Acids in Both Molten and Solvent Systems

Mohammad Reza Razzaghi, Luis Trevisi, Ana Carolina Jerdy, Dai Phat Bui, Steven Crossley, Lance Lobban

Multilayer packaging films often use polyethylene (PE) as the major polymer and polyvinyl alcohol-co-ethylene (EVOH) as one of the layers for its oxygen-barrier properties. Physical recycling of these films is challenging since melts of the film show poor reusability due to polymer segregation. One option to address this problem is selective conversion of EVOH via dehydration under mild acid catalytic conditions, followed by hydrogenation of the resulting carbon-carbon double bonds to enhance compatibility with PE.

In this study we investigate the use of zeolites and other solid acid catalysts to carry out the first step, dehydration of PVA. Not surprisingly, in the molten phase dehydration rates are strongly influenced by catalyst pore size and acid site density indicating strong mass transfer effects. In order to quantify the extent of mass transfer limitations, the catalysts were characterized using IPA-TPD as well as FTIR measurement of adsorbed pyridine and di-tert-butylpyridine to quantify total acid sites and external acid sites, assuming that the di-tert-butylpyridine is able to access sites only in the pore mouths. Experimental rates of EVOH dehydration were then used to determine separately the rates on internal and external acid sites. Comparison of these rates with rates measured in solvent systems was used to further elucidate external mass transfer rates in these complex polymer systems. This approach offers valuable insights for optimizing acid catalysts for recycling multilayer films containing PVA, thus advancing sustainability in polymer recycling processes.

385am Continuous Catalytic Upcycling of Multilayered Plastic Packaging in a Biphasic System

Dai Phat Bui, Laura Alejandra Gomez Gomez, Samira Abdolbaghi, Lance Lobban, Steven Crossley

Multilayered plastic packaging films often contain materials such as blends of polyethylene (PE) with oxygen barrier polymers such as poly(vinyl alcohol-co-ethylene) (EVOH). These materials when layered provide excellent barrier performance as well as stability, leading to their wide application in food and pharmaceutical industries. However, its barrier performance and the inherent immisciblity of the two polymers makes recycling challenging, resulting in waste in landfills. Melting or mechanical recycling produces low value of mixed plastics, while chemical reactions in single solvents are often unfavorable due to solubility issues with the various phases. In this work, we use a γ-Valerolactone – Decalin biphasic solvent system that allows the Pd supported on TiO_2 nanotubes to selectively deoxygenate polar layers without cracking the hydrocarbon backbone in one phase while isolating the nonpolar products in another phase. We show how the active metal-support interface selectively cleaves C-O bonds to convert waste EVOH impurities in multicomponent films into a compatible PE stream. Further, the GVL solvent remains in the same reaction mixture allowing for continuous separation and removal of the pure resulting polyolefin stream. The biphasic solvent system shows significant potential for selective upgrading and separation of complex plastic waste streams in a single pot. Conversion levels achieved in a biphasic system are far superior to those observed in a monophasic solvent system due to challenges associated with dissolving multilayered plastic packaging, prolonging reactions, and separating products in the latter. Furthermore, the reaction kinetics over the catalysts were studied supporting further studies. Through deoxygenation and continuous removal of products through in a single reaction vessel, this work addresses one of the important limitations of multilayered plastic packaging conversion.

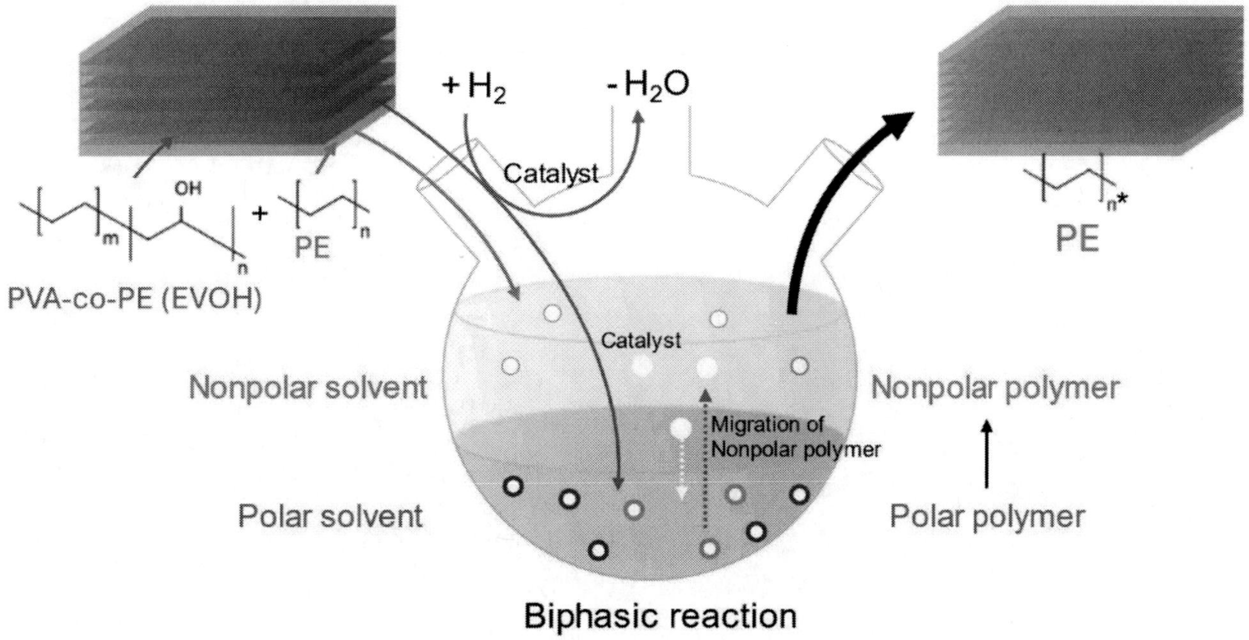

385an Functional Upcycling of Polyurethane Thermosets into Value-Added Thermoplastics Via Small-Molecule Carbamate-Assisted Decrosslinking Extrusion

Jared Nettles, Kailong Jin, Timothy E. Long

Polyurethanes (PUs) deliver exceptional, versatile properties and find a home in most aerospace, insulation, and comfort applications. The thermoset nature of many commercial PUs, however, prohibit common melt reprocessing methodologies. Kinetic and thermodynamic limitations fundamental to the urethane linkage hinder the wholesale recovery of viable PU monomers from solvolytic depolymerization processes. Dynamic carbamate exchange empowered by reactive extrusion emerges as a novel reprocessing methodology to transform PU thermosets into reprocessable covalent adaptable networks (CANs). This approach, however, has been limited to thermoset-to-thermoset reprocessing, which requires high reaction temperatures due to their well preserved, viscous network structure. This study reports the development of a reactive extrusion decrosslinking process which converts PU networks into thermoplastics in a rapid, solvent-free manner. Low molecular weight, carbamate-functionalized small molecules simultaneously decrosslink the PU network while functionalizing the thermoplastic in the melt state. Amendable to both crosslinked PU films and foams, the process is generalizable towards multiple photoactive functionalities, with chain-end functionalization confirmed by analyzing purified extrudates after dialysis. We seek to propose the decrosslinking technique as a platform to transform crosslinked PU waste into a library of thermoplastic PUs for diverse applications, such as adhesives, photoresins, and stimuli-responsive materials. Altogether, this process represents a powerful reaction extrusion mediated decrosslinking methodology applicable towards other commercial thermosets capable of dynamic exchange reactions, such as polyurea and polyester networks.

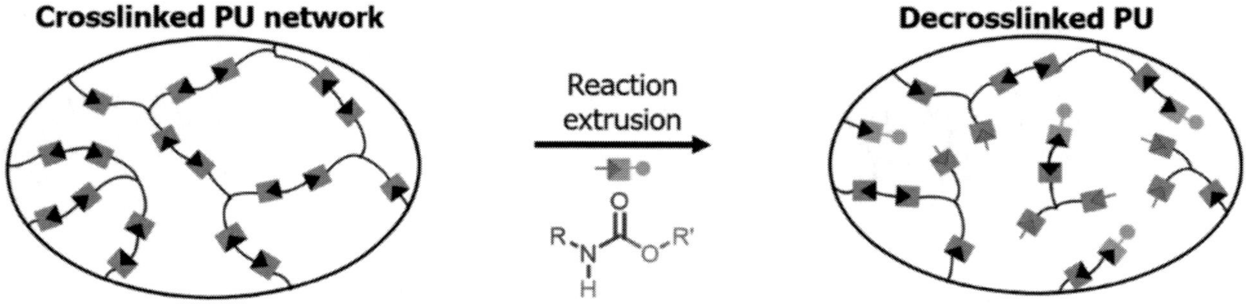

385ao Recycling of Post-Consumer Polyethylene Terephthalate (PET) Waste Via Glycolysis Using Ionic Liquid Catalyst

Maria Coleman, Fahimeh Forouzeshfar, Joseph G. Lawrence

The mechanical recycling of complex polyethylene terephthalate (PET) waste, including multilayer packaging, colored plastics, and mixed plastic waste streams, presents significant challenges due to the presence of contaminants, and additives that can compromise the quality and properties of recycled material. In response, advanced chemical recycling methods such as glycolysis has emerged as a promising solution to address the limitations of mechanical recycling. This study focused on the glycolysis of post-consumer PET waste with using an ionic liquid (IL) catalyst containing 1,8-diazabicyclo[5.4.0]undec-7-ene (DBU) and oxalic acid (OA). The research initially examined the impact of various reaction parameters, including temperature, time, catalyst dosage, and the ethylene glycol (EG) to PET ratio, on the conversion of PET and the yield of the monomer bis(2-hydroxyethyl) terephthalate (BHET). Under optimized conditions, the study evaluated the glycolysis performance of various PET waste, including textiles, bottle, and colored PET containers. Additionally, the study focused on separating additives and pigments from the reaction mixture to enhance the quality of resulting BHET for potential PET synthesis. The study also proposed a reaction mechanism to explain the glycolysis process. Furthermore, a techno-economic analysis was conducted to evaluate the feasibility and economic viability of scaling up the glycolysis process for industrial application.

385ap Carbon-Efficient Upcycling of Plastic Wastes to Biodegradable Polymers Enabled By CO_2 Plasma

Caixia Wan, Krishanthi Mapa Mudiyanselage, Xianglan Bai, Harish Radhakrishnan, Sultan Ul Iffat Uday

This work aims at upcyling of plastic wastes into polyhydroxyalkanoic acids (PHAs) enabled by CO_2 plasma-assisted oxidative plastic deconstruction. It provides a viable solution by supplying plastic-derived substrates for microbial utilization due to significantly reduced molecular weight and various oxygen-containing functionalities (e.g., organic acids, fatty alcohols). Pseudomonas strains are used to convert plastic deconstruction intermediates to synthesize medium chain length PHAs (mcl-PHAs). The plastic-derived substrates contain no inhibitory compounds and well support microbial growth and PHA accumulation. Carbon conversion efficiency is also significantly improved by optimizing operating parameters (e.g., pH, aeration, and feeding strategy) as well as involving sequestering CO_2 generated from fermentation.

619a Catalytic Cracking of Crude Waste Plastic Pyrolysis Oil to Maximize Light Olefin Production in a Pilot-Scale Circulating Fluidized Bed Reactor

Xuan Tin Tran, Dae Hun Mun, Jiho Shin, Na Young Kang, Yong-Ki Park, Do Kyoung Kim

The invention and widespread adoption of plastic have been considered as one of the most significant advancements of the 20th century, exerting a profound impact on human life. However, alongside the numerous benefits offered by plastic products, the indiscriminate use of plastic has emerged as a considerable challenge that necessitates prompt attention. For instance, out of the 353 million metric tons (MMT) of plastic waste generated in 2019, only 9% underwent recycling, while the majority ended up in landfills (50%), incineration (19%), and mismanagement (22%) [1]. In addition to mismanaged plastic waste, the handling of plastic waste through landfilling and incineration has led to considerable environmental pollution and ecological hazards. Moreover, enormous amounts of greenhouse gases are generated during the production and disposal of plastics [2]. Therefore, recycling of waste plastics with high efficiency is indispensable for preventing environmental pollution as well as low carbon emission, leading to a climate-neutral future.

Among various waste plastic recycling methods, mechanical recycling and thermochemical recycling have been highlighted as the most promising technologies [2]. Mechanical recycling stands out as one of the most energy-efficient and sustainable options. However, it demands highly sorted and clean waste plastics; otherwise, it frequently yields lower-quality products compared to those derived from virgin polymers due to the inherent impurities in plastic waste streams. On the other hand, thermochemical recycling converts plastic waste into its monomer, and low-quality or mixed plastic waste can be used. From a different perspective, among the various types of plastic, polyolefinic plastics, such as polyethylene (PE) and polypropylene (PP), which are widely used as packaging materials, account for more than 60% of the global waste plastic production owing to their short product lifetime [3]. Therefore, there is an upmost emphasis on developing new thermochemical recycling technology to produce light olefins (e.g. ethylene, propylene, butenes) from contaminated and mixed waste polyolefins, suitable for the their remanufacturing.

In this context, ongoing endeavors are being pursued to advance the commercialization of this recycling pathway [4], as illustrated in Figure. This approach involves subjecting waste plastics to pyrolysis in an oxygen-free environment, resulting in the generation of gas, liquid oil, and residue. The Crude waste plastic pyrolysis oil (WPPO) is subjected to post-treatment techniques, such as hydroprocessing, to remove contaminants and impurities, leading to production of t-WPPO (hydrotreated WPPO). Subsequently, the naphtha range of t-WPPO is converted into light olefins utilizing well-established naphtha-steam-cracking units.

However, the state-of-the-art commercial scheme is thought to have several drawbacks, mostly derived from the notable differences in properties between conventional naphtha and WPPO. First, crude WPPO typically comprises a broader range of carbon numbers (from C_5 to C_{44}) than conventional naphtha, which contains paraffinic straight-chain hydrocarbons from C_5 to C_9 [5]. Owing to this difference, only a fraction of the pyrolysis oil is utilized, or the hydrocarbon chain length requires further reduction to produce a steam-cracker-suitable feedstock. Additionally, crude WPPO typically contains a substantial amount of olefins (30–50 wt%). Industries typically limit the olefin content of the feedstocks for thermal naphtha cracking to a maximum of 2 wt% owing to the tendency of high olefin contents to cause coke formation [5]. This imbalance could be resolved by moderately diluting crude WPPO with fossil naphtha or subjecting it to a post-saturation process to convert olefinic compounds into single-bond ones for meeting the given specifications. Finally, WPPO contains various kind of contaminants such as heteroatomic compounds (nitrogen, sulfur, oxygen and chlorine) and metallic compounds, some of which exceed typical specification of naphtha steam cracker [5]. In addition to the partial dilution of WPPO with conventional naphtha, catalytic hyprotreating such as hydrodesulfurization (HDS), hydrodenitrogenation (HDN), hydrodeoxygenation (HDO), hydrodechlorination (HDCl), Hydrodemetalation (HDM), which are widely used in commercial application, are thought to be useful for removing heteroatomic and metallic compounds in WPPO to meet the current contaminants specifications set for industrial steam cracker feedstocks. However, diluting WPPO with naphtha restricts its use, whereas post-saturation requires an appreciable amount of hydrogen, which can be detrimental to economic feasibility.

In our previous work, a new naphtha catalytic cracking process based on a circulating fluidized bed reactor (CFBR) was developed collaboratively by our group, SK Corporation, and Kellogg Brown & Root (KBR) to replace the traditional naphtha thermal cracking process, which was successfully commercialized under names ACO^{TM} (Advanced Catalytic Olefins) and later K-COT^{TM} (KBR catalytic olefins Technology) [6]. This system presents several advantages over conventional non-catalytic naphtha thermal cracking, including enhanced feedstock flexibility (the utilization of heavier feedstocks, olefin-rich feed streams, and oxygenates), higher overall yields of light olefins (approximately 10–25% more total olefins than a typical steam cracker), a higher propylene to ethylene ratio (1:1 propylene-to-ethylene ratio compared to 1:2 ratio delivered by steam crackers), and reduced energy consumption due to relatively lower cracking temperature (reduced from 850 °C to below 700 °C).

Given the distinctive properties of WPPO and the advantageous nature of catalytic cracking in CFBR systems, particularly in processing feeds with such properties, catalytic cracking of WPPO within a CFBR system emerges as a highly promising alternative to conventional thermal cracking, akin to the case of naphtha. This system could not only address the inherent limitations associated with using WPPO but also offers notable advantages such as reduced energy consumption and increased yields of light olefins.

In this regard, we investigated the feasibility of continuous catalytic cracking of crude WPPO in a microactivity test (MAT) unit [7]. Interestingly, the catalytic WPPO cracking led to a remarkable increase in C_2–C_4 olefins yield, surpassing those obtained using catalytic naphtha cracking as well as thermal WPPO cracking and thermal naphtha cracking. This superiority of WPPO as a feedstock over naphtha was attributed to its unique properties, particularly its high content of long-chain paraffins and olefins, which facilitated enhanced crackability on the catalyst compared to the predominantly short-chain paraffins found in conventional naphtha. Additionally, the feasibility of continuous operation was examined by monitoring the tendencies of the catalyst to deactivate rapidly via coke formation and to recover its activity through regeneration. Also, the continuous catalytic cracking of crude WPPO for light olefin production was assessed in a pilot-scale CFBR (1,000 g of WPPO input/hour scale). A comprehensive investigation was carried out to examine the effect of operating parameters on the product distribution, and a 24-hour durability test was employed to scrutinize the stability of the CFBR system throughout continuous catalytic cracking of crude WPPO.

The findings of this study can be exploited to implement a new approach to directly produce light olefins by integrating the catalytic cracking scheme for WPPO in a circulating fluidized bed reactor system, as shown in Figure. This strategy offers several advantages over most commercially developed approaches, such as permitting the utilization of the entire crude WPPO range without requiring any post-treatment processes. This can lead to high light-olefin yields, continuous coke removal, and reductions in the energy consumption. Moreover, this approach eliminates the need for hydrogen, which is typically required during the post-hydroprocessing stage of commercial schemes.

[1] Global plastics outlook: economic drivers, environmental impacts and policy options 2022.

[2] Vollmer I, Jenks MJF, Roelands MCP, White RJ, van Harmelen T, de Wild P, et al. Beyond mechanical recycling: giving new life to plastic waste. Angewandte Chemie - International Edition 2020;59:15402–23.

[3] Yang RX, Jan K, Chen CT, Chen WT, Wu KCW. Thermochemical conversion of plastic waste into fuels, chemicals, and value-added materials: a critical review and outlooks. ChemSusChem 2022;15:e202200171.

[4] Li H, Aguirre-Villegas HA, Allen RD, Bai X, Benson CH, Beckham GT, et al. Expanding plastics

recycling technologies: chemical aspects, technology status and challenges. Green Chemistry 2022;24:8899–9002.

[5] Kusenberg M, Eschenbacher A, Djokic MR, Zayoud A, Ragaert K, De Meester S, et al. Opportunities and challenges for the application of post-consumer plastic waste pyrolysis oils as steam cracker feedstocks: To decontaminate or not to decontaminate? Waste Management 2022;138:83–115.

[6] Curtis Eng, P. Ramamurthy, Shin C. Kang, Sun Choi, S.H.Oh, Y. K. Park. A catalytic cracking process for ethylene and propylene from paraffin streams-The advanced catalytic olefins (ACO) process. AIChE spring national meeting, Houston, Texas: 2007.

[7] Tran XT, Mun DH, Shin J, Kang NY, Park DS, Park YK, et al. Maximizing light olefin production via one-pot catalytic cracking of crude waste plastic pyrolysis oil. Fuel 2024;361.

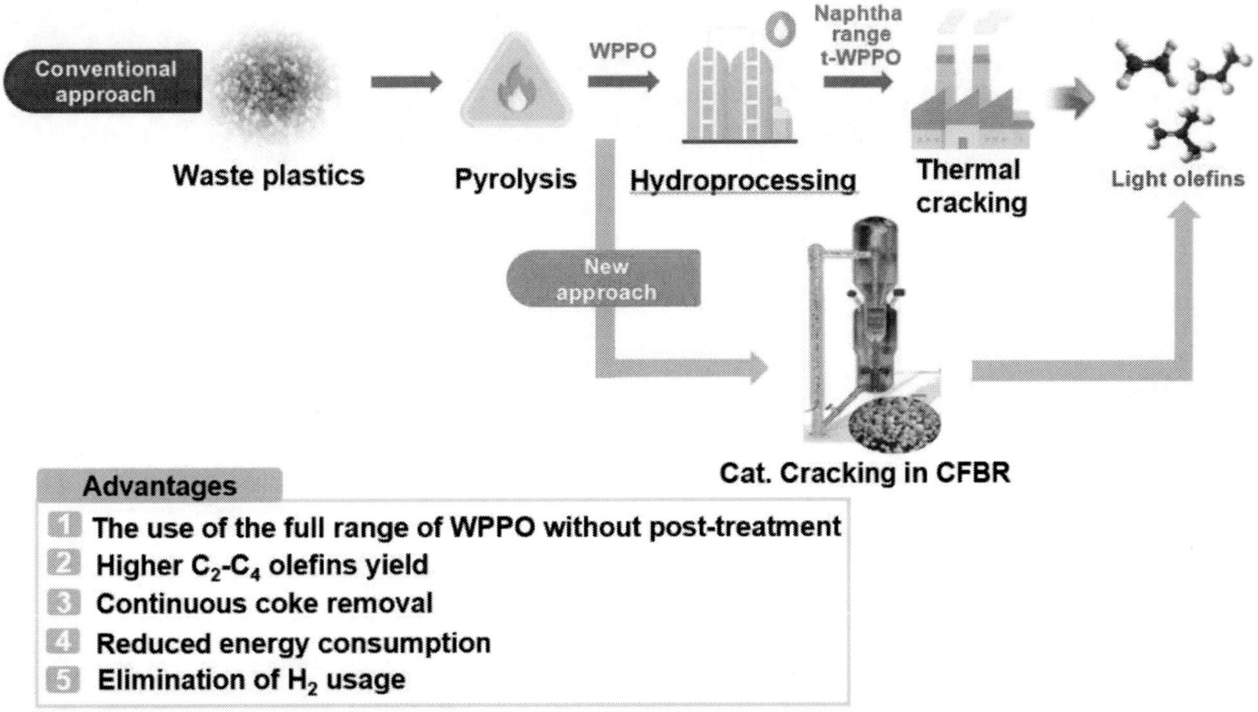

619b End-of-Life Plastic Liquification: Opportunities and Challenges of Scaled Continuous Recycling of Molecules.

Sheldon Cotts

A core challenge of technological realization is the translation of scale. Alterra's patented technology for converting mixed discarded plastic into synthetic crude-like oil has successfully scaled from bench to commercial scale with a current capability of > 50 tons of feedstock per day. Leveraging an indirectly heated rotary reactor enables high yielding hydrocarbon capture while subsequently isolating colorants, fillers, and contaminates. However, a current pinch point in the market is the wide variability of discarded plastic streams and a lack of feedstock commercialization. This presentation will outline how our process has matured to overcome technical barriers, and how feedstock sourcing is commercially commoditized.

619c Microwave-Assisted Catalytic Plastic Upcycling

Chunlin Luo, Brandon Robinson, Jianli Hu, Yuxin Wang

Microwave-Assisted Catalytic Plastic Upcycling

Chunlin Luo, Brandon Robinson, Jianli (John) Hu, Yuxin Wang*

Chemical and Biomedical Engineering Department, West Virginia University

yuxin.wang1@mail.wvu.edu

The chemical upcycling of plastic represents an attractive approach of using plastic waste as feedstock in promoting the circular economy and decarbonizing the chemical industry. Currently, the majority of plastic ends up in landfills or the environment, harming the ecosystem and affecting the natural environment. Thermochemical pathways (e.g., pyrolysis, gasification) and high pressure hydrogenolysis enable depolymerization films to produce small molecules that could be used as fuel or integrated into chemical refineries. However, these thermochemical processes suffer from severe challenges which include inevitably higher energy input leading to higher greenhouse gas (GHG) emission (i.e., ~12 kgCO_2/1 kg H_2 production in gasification); poor product selectivity; low value of products, thus requiring significant upgrading for converting to valuable products.

In this work, we reported the application of microwave catalysis technology application in plastic upcycling. Microwave heating has many advantages over conventional heating, including non-contact, selective and rapid heating, and quick start-up and shutdown. This fundamental difference results in important advantages in using microwave to initiate the catalytic depolymerization of plastics, particularly, the microwave irradiation's selectivity, as the microwave almost exclusively interact with the absorbing catalyst particles whilst the plastic substrate remains cold; thus, the self-decomposition of plastic polymers and many side reactions that occur at high temperatures are effectively limited.

The microwave catalysis technology demonstrated several advantages in plastic upcycling: (1) lower energy consumption leading to lower GHG emission, < 300 °C under atmospheric pressure; (2) higher selectivity of target product; (3) higher value of products (e.g., BTX aromatics (benzene, toluene, and xylene), ethylene, carbon nanofibers, and H_2, etc.); (4) moreover, the product can be adjusted based market requirement. For example, a high BTX aromatics (benzene, toluene, and xylene) yield of 44.7 % from low-density polyethylene (LDPE) was achieved using ZSM-5 catalyst at a low temperature of 300 °C under atmospheric pressure. Also, as shown in Figure 1, the high-quality separated CNFs can be generated from LDPE by microwave catalysis, meanwhile, the high purity H_2 (>93%) generated.

This proof-of-concept heralds an exciting new era of applications for the responsible and sustainable recycling of waste plastic, in which plastic are efficiently converted back to high-value petrochemical products.

619d Screening Green Solvents for Multilayer Plastic Film Separation

Ugochukwu Ikegwu, Victor M. Zavala, Reid Van Lehn

The increasing global demand for specialized product packaging materials beyond the capabilities of single-layer plastics has driven the development of multilayer plastic films. These multilayer plastic films often consist of two to several polymers, including Polyethylene (PE), Polyethylene Terephthalate (PET), Ethylene Vinyl Alcohol (EVOH), and Polyvinyl Chloride (PVC), with each polymer offering diverse functions towards enhancing product safety, shelf life, and handling (Alias et al. 2022; Barry, 2022). However, the intricate mix of these polymers increases the difficulty in recycling multilayer plastic films via mechanical processes, leaving their end-of-life fate to landfills and incineration (Hutchings et al. 2021; Milbrandt et al. 2022; Rudolph et al. 2020). Solvent-based separation technologies are promising for re/up-cycling these multilayer plastic films (Zhao et al. 2018). A manifestation of this approach is the so-called Solvent Targeted Recovery And Precipitation (STRAPTM) process, which is a technology that sequentially separates polymer components of multilayer plastic films through a series of solvent washes using selective solvents (Walker et al. 2020). The STRAPTM process has exhibited potential in the recovery of high-quality resins from multilayer plastic film wastes (Zhao et al. 2018) and offers environmental (Munguía-López et al. 2023) and economic (Sánchez-Rivera at al. 2021) advantages over fossil-based plastic films. A critical step in the actualization of the STRAP™ process involves the identification of solvents capable of selectively dissolving a polymer layer while leaving the other layers undissolved. Molecular-scale models have been harnessed to predict the temperature-dependent solubilities of diverse polymers in organic solvents (Zhou et al. 2021). These models have been integrated with experimentally derived solubility data to improve the accuracy of polymer solubility predictions on a large scale across ~1000 solvents (Zhou et al. 2023). Although the STRAP™ process shows promise for revolutionizing the plastic recycling sector, significant challenges persist. While solvents typically remain unconsumed in solvent-mediated recycling technologies, repeated usage can lead to minute quantities escaping into the environment or their intentional disposal, potentially generating new waste streams. The ecological impact of such waste streams can be significant if the solvents in question possess toxic properties. Moreover, like other solvent-mediated processes, there is a heightened risk of prolonged exposure to toxic solvents among operational personnel. Therefore, beyond prioritizing selective solubility, it is critical to assess the greenness of solvents during screening, contributing to a more sustainable process design – an essential facet of green chemistry (Gao et al. 2020).

This study introduces a multistep green solvent selection framework for separating multilayer plastic films into their constituent resins. This framework, comprising eight sequential steps, focuses on evaluating polymer selective solubility (steps 1-5), solvent greenness (steps 6-7), and cost of greenness (step 8). This framework begins with employing molecular scale models to quickly predict temperature-dependent solubilities of the multilayer plastic film components (i.e., polymers) in ~1000 solvents. Utilizing short oligomers to represent polymers, molecular dynamic (MD) simulations, and the COnductor-like Screening MOdel for Real Solvents (COSMO-RS), grounded in quantum chemistry principles, enables accurate solubility predictions. Subsequently, a fast solvent screening process is employed to identify solvents capable of selectively dissolving and precipitating a target polymer under standard atmospheric pressure. To assess the greenness of the selected solvents, various criteria, including the operational energy required for the dissolution process, octanol-water partition coefficients (LogP), and greenness levels of solvents using the industrial solvent selection guidelines provided by GlaxoSmithKline (GSK), are employed. However, these guidelines do not contain information about most of the solvents in our study. Therefore, solvent regulatory standards such as the Registration, Evaluation, Authorization, and Restriction of Chemical Regulation (REACH) initiative and the US Environmental Protection Agency (EPA) are consulted to identify the potential hazards and any regulatory restrictions associated with each solvent on their industrial usage. The cost of greenness is evaluated by performing a techno-economic and life cycle analysis on the STRAP™ process using each of the candidate solvents to explore the tradeoffs between solubility, cost, and solvent greenness, utilizing the open-source software BioSTEAM (Cortes-Pena et al. 2020), and OpenLCA (Ciroth et al. 2020). Subsequently, this framework is applied to an industrial multilayer plastic film containing PE, EVOH, and PET to identify green solvents for its separation and associated tradeoffs.

Alias, A. R., Wan, M. K., & Sarbon, N. M. (2022). Emerging materials and technologies of multilayer film for food packaging application: A review. *Food Control,* 136, 108875.

Barry, M. A. (2022). The Science and Technology of Flexible Packaging: Multilayer Films from Resin and Process to End Use. *Elsevier.*

Ciroth, A., Di Noi, C., Lohse, T., & Srocka, M. (2020). openLCA 1.10. *GreenDelta GmbH. Available online at https://www.openlca.org/wp-content/uploads/2020/02/openLCA_1.10_User-Manual.pdf.*

Cortes-Pena, Y., Kumar, D., Singh, V., & Guest, J. S. (2020). BioSTEAM: a fast and flexible platform for the design, simulation, and techno-economic analysis of biorefineries under uncertainty. *ACS sustainable chemistry & engineering*, *8*(8), 3302-3310.

del Carmen Munguía-López, A., Göreke, D., Sánchez-Rivera, K. L., Aguirre-Villegas, H. A., Avraamidou, S., Huber, G. W., & Zavala, V. M. (2023). Quantifying the environmental benefits of a solvent-based separation process for multilayer plastic films. *Green Chemistry*, *25*(4), 1611-1625.

Gao, F., Bai, R., Ferlin, F., Vaccaro, L., Li, M., & Gu, Y. (2020). Replacement strategies for non-green dipolar aprotic solvents. *Green Chemistry*, *22*(19), 6240-6257.

Hutchings, N., Smyth, B., Cunningham, E., Yousif, M. & Mangwandi, C. (2021). Comparative life cycle analysis of a biodegradable multilayer film and a conventional multilayer film for fresh meat modified atmosphere packaging–and effectively accounting for shelf-life. *Journal of Cleaner Production,* 327**,** 129423.

Milbrandt, A., Coney, K., Badgett, A. & Beckham, G. T. (2022). Quantification and evaluation of plastic waste in the United States. *Resources, Conservation and Recycling,* 183**,** 106363.

Rudolph, N., Kiesel, R., & Aumnate, C. (2020). *Understanding plastics recycling: Economic, ecological, and technical aspects of plastic waste handling.* Carl Hanser Verlag GmbH Co KG.

Sánchez-Rivera, K. L., Zhou, P., Kim, M. S., González Chávez, L. D., Grey, S., Nelson, K., ... & Huber, G. W. (2021). Reducing antisolvent use in the STRAP process by enabling a temperature-controlled polymer dissolution and precipitation for the recycling of multilayer plastic films. *ChemSusChem*, *14*(19), 4317-4329.

Walker, T. W., Frelka, N., Shen, Z., Chew, A. K., Banick, J., Grey, S., ... & Huber, G. W. (2020). Recycling of multilayer plastic packaging materials by solvent-targeted recovery and precipitation. *Science advances*, *6*(47), eaba7599.

Zhao, Y. B., Lv, X. D., & Ni, H. G. (2018). Solvent-based separation and recycling of waste plastics: A review. *Chemosphere,* 209, 707–720.

Zhou, P., Sánchez-Rivera, K. L., Huber, G. W., & Van Lehn, R. C. (2021). Computational Approach for Rapidly Predicting Temperature-Dependent Polymer Solubilities Using Molecular-Scale Models. *ChemSusChem*, *14*(19), 4307-4316.

Zhou, P., Yu, J., Sánchez-Rivera, K. L., Huber, G. W., & Van Lehn, R. C. (2023). Large-scale computational polymer solubility predictions and applications to dissolution-based plastic recycling. *Green Chemistry*, *25*(11), 4402-4414

619e Evaluation and Comparison of PET Depolymerization Routes

Adam McNeeley, Y. Liu

Poly(ethylene terephthalate) (PET) is the most widely produced thermoplastic and is commonly used in consumer products such as bottles, packaging and textiles. There is potential economic incentive to recycle commercial PET waste because of its relatively high value and the ability to collect waste with high contents of PET. Collection infrastructure and mechanical recycling of PET bottles is well developed, but mechanical recycling is limited in its ability to purify waste PET to a virgin quality. Chemical recycling of PET offers a way to expand the scope of PET waste materials that can be recycled and reduce cleaning and sorting requirements of reclaimers. Chemical recycling involves the conversion of PET into relevant monomers conventionally used to produce PET and provides the advantage of enabling separation and purification processes such as crystallization and distillation, unavailable to polymers. This allows the monomers to be purified such that they can be polymerized into virgin quality PET.

The reversible nature of the ester linkages of PET means that the polymer can be converted into many different monomers. Our work investigates process modeling of the three primary routes to depolymerize PET: methanolysis, glycolysis, and hydrolysis. All three depolymerization routes are being developed for application on a commercial scale, but there is a lack of understanding of the differences among the depolymerization routes. We first compare the physical and thermodynamic properties of the products and reactants of the different routes and demonstrate how these properties lead to fundamentally different operating characteristics in the reaction and purification sections of the process. We then cover in detail the purification methods described in patents by different industrial technologies and demonstrate similarities and differences in the purification strategies of different depolymerization routes. We use our patent survey of the industrial depolymerization technologies to present a general framework for a PET depolymerization process.

The final part of this work involves the modeling of the different PET depolymerization processes using Aspen Plus. We build detailed models assuming common impurities and byproducts and simulate major unit operations. Process simulation is developed by integrating industrial literature review, knowledge of process design and heat integration. We present process innovations using advanced heat integration and process intensification techniques for each technology. During process evaluation, we consider uncertainties regarding optimal solvent rates for the reaction and purification sections and demonstrate the relationship between these variables and process energy demand.

From this investigation, we draw general conclusions about the advantages and disadvantages offered by each depolymerization route. Overall, the glycolysis process is the simplest and least energy-intensive process; however, challenges with isolating its monomeric product from oligomers and other heavy impurities make it less attractive for processing low-purity waste PET. Methanolysis has the potential to be competitive with glycolysis processes in terms of energy demand and the monomer dimethyl terephthalate (DMT) has a high enough volatility to be more easily purified by evaporation or distillation. Methanolysis is disadvantaged because DMT is used as a feed in only a few PET existing polymerization units. An extra reaction step following the purification of dimethyl terephthalate (DMT) by reacting with water to form terephthalic acid (TPA) may be necessary. The hydrolysis process is the least attractive of the depolymerization routes because of the necessity to convert the product to a carboxylate salt, which consumes an acid and base, the large amount of water to dissolve the carboxylate salt, and the added complexity handling a salt byproduct and separating ethylene glycol. We expect that a continuous, properly heat- integrated PET depolymerization process to have a total energy demand between 6,000-10,000 kJ per kg of PET.

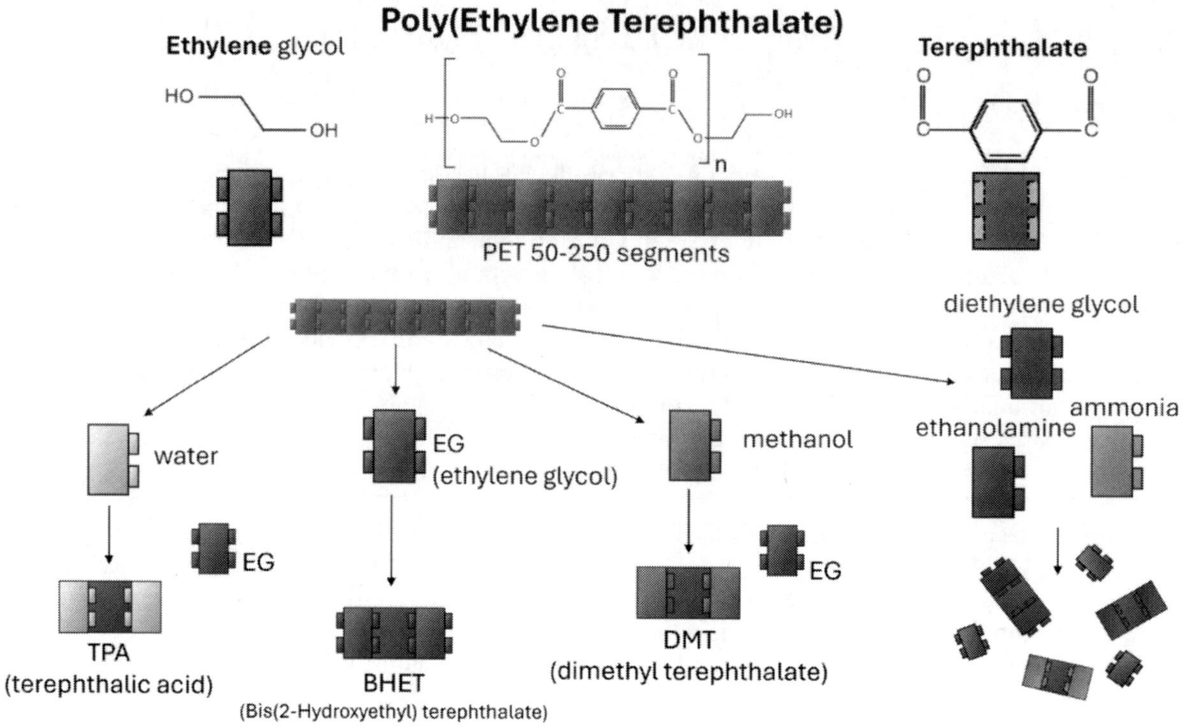
Poly(Ethylene Terephthalate)
Ethylene glycol
HO
OH
Terephthalate
PET 50-250 segments
diethylene glycol
ethanolamine
ammonia
water
EG
(ethylene glycol)
methanol
EG
EG
EG
TPA
(terephthalic acid)
BHET
(Bis(2-Hydroxyethyl) terephthalate)
DMT
(dimethyl terephthalate)

619f Calculated Physicochemical Properties of Glycerol-Derived Solvents to Drive Plastic Waste Recycling

Ademola Soyemi, Tibor Szilvasi

Plastics serve modern life in several ways, but their extensive use and mismanagement of the accompanying waste contributes to the accumulation of plastic waste in landfills or in the natural environment. A promising method of combating plastic waste is solvent-based recycling which uses a solvent to selectively dissolve plastic material which can then be recovered and reprocessed into upcycled products. In this work, we propose the functionalization of glycerol, a natural product from biomass as well as a byproduct of biodiesel production, that can produce new and potentially less harmful solvents capable of dissolving plastic waste while also overcoming the health and environmental hazards of commonly used solvents. In this work, we have developed a database of ~10,000 glycerol-based solvents and calculated their physical properties such as the dipole moment, density, melting point, LogS, LogP, and Hansen Solubility Parameters (HSPs) using a combination of Density Functional Theory, machine learning and Hansen theory. We show that the physical properties can be tuned based on the attached functional groups and based on the HSPs of the solvents and selected plastics in addition to the melting point of the solvents, we give recommendations for solvents that can carry out selective plastic dissolution for a complex plastic waste feedstock (*Ind. Eng. Chem. Res.*, 62, 6322-6337 2023. DOI: 10.1021/acs.iecr.2c04567). We also highlight successful experimental verification, for example the selective dissolution of polystyrene in glycerol-derived solvents at room temperature.

619h Chemical Recycling of Plastic Wastes Using Hydrogenolysis Method

Emmanuel Dada, Temitope Esan, Carol Akpan

Plastic waste constitutes a global waste problem in the environment and landfills. To mitigate this problem, a transition is needed from the concept of single-use and discard to a more environmentally beneficial circular economy whereby the plastic waste is converted to monomers to produce new or the same plastics, and other useful chemicals. The objective of this project is to develop an energy-efficient and low-temperature chemical recycling of waste plastics that will effectively cleave the C-C bonds in polyolefins. This will be carried out using the hydrogenolysis method which uses hydrogen to depolymerize the mixed plastic wastes by developing a highly efficient reactor with a low-cost catalyst that has high selectivity to the desired monomers of interest and/or other useful chemicals. This can be achieved by careful selection of catalysts that will give a high-value product. With more than 5 billion metric tons of global plastic waste generated annually, the impact of using hydrogenolysis method to chemically depolymerize mixed plastic wastes and turn them into monomers and/or other useful chemicals has a great impact on solving the global environmental pollution problem and enhancing the circular economy of plastic wastes.

AUTHOR INDEX

AIChE
120 Wall Street, FL 23
New York, NY 10005-4020

ISBN 979-8-3313-1672-3

CURRAN ASSOCIATES INC.
proceedings.com